Organic Agriculture and Climate Change Mitigation

A Report of the Round Table on Organic Agriculture and Climate Change

Urs Niggli
Nadia El-Hage Scialabha

United Book Prints

1st Reprint in India, 2015

Published by:

United Book Prints
4806/24, Ansari Road,
Daryaganj,
New Delhi - 110 002
Tel.: 011-41511055

ISBN: 978-93-83692-06-4

Printed in India

CONTENTS

LIST OF FIGURES

LIST OF TABLES AND BOXES

ABBREVIATIONS AND ACRONYMS

ANOVA	Analysis of variance
BSI	British Standards Institution
CDM	Clean Development mechanism
CEDECO	Educative Cooperation for the Development of Costa Rica
CERS	Certified Emission Reductions
CMA	Comprehensive meta-Analysis
DEFRA	Department for Environment, Food and Rural Affairs
DOK	Biologisch-dynamisch, Organisch-biologisch and Konventionell
EF	Emission Factors
EU	European Union
FADN	Farm Accountancy Data Network
FAO	Food and Agriculture Organization of the United Nations
FIBL	Research Institute of Organic Agriculture
GHG	Greenhouse Gas
ICEA	Environmental and Ethical Certification Institute
ICROFS	International Centre for Research in Organic Food Systems (Denmark)
IFA	International Fertilizer Association
IFOAM	International Federation of Organic Agriculture movements
IGO	Intergovernmental organizations
ILCD	International Reference Life Cycle Data System

ISO	International Organization for Standards
LCA	Life Cycle Assessment
MRV	Monitoring Reporting and Verification
NAMA	Nationally Appropriate mitigation Actions
NAPA	National Adaptation Programmes of Action
OFA	Organic Federation of Australia
OTC	Over-the-Counter (Carbon market)
PAS	Publicly Available Specification
PLFA	Phospholipid Fatty Acids
REDD	Reducing Emissions from Deforestation and Forest Degradation in developing countries
RTOACC	Roundtable on Organic Agriculture and Climate Change
SALM	Sustainable Agricultural Land management
SEAE	Spanish Society for Organic Farming
SOC	Soil Organic Carbon
SPC	South Pole Carbon Asset management Ltd
UNFCCC	United Nations Framework Convention on Climate Change
VCM	Voluntary Carbon market

PREFACE

During the next decades, billions of people, particularly those in developing countries, will face changes in climate patterns that will contribute to severe water shortages or flooding, and rising temperatures that will cause shifts in crop growing seasons. This will increase food shortages and distribution of disease vectors, putting populations at greater health and life risks. The predicted temperature rise of 1 to 2.5°C by 2030 will have serious effects, especially in terms of reduced crop yield. The productivity of farms is likely to diminish because of climate change, especially in the 40 poorest countries in Africa and Asia. Increased drought periods in many parts of the world and erratic rainfalls will endanger yield stability and put global food production at risk.

As the world seeks solutions for facing the reality of changing climates, the importance of mitigating the effects of greenhouse gas (GHG) emissions becomes increasingly significant, especially in the agriculture sector which both emits and sequesters GHGs. Agriculture causes approximately one-third of global GHGs when direct energy use, emissions from livestock, the production of fertilizers, pesticides, machinery and equipment as well as soil degradation and land-use change for feed production are taken into account.

yet, agriculture and, in particular, organic agriculture can be part of the solution to mitigate GHG gases through farming practices that build soil fertility, avoid use of synthetic fertilizer and improve carbon sequestration. The report of the Intergovernmental Panel on Climate Change (IPCC) on the role of agriculture considers many techniques packed into organic management as relevant mitigation and adaption actions, such as the integration of leguminous plants into the crop rotations, excellent soil cover, mixed farming systems and the longevity of ruminants. The Round Table on Organic Agriculture and Climate Change (RTOACC) is a newly launched initiative dedicated to increasing understanding and quantifying the role that organic agriculture can play in climate change mitigation and adaptation – in addition to its already understood contribution in areas such as reducing use of chemical pesticides and biodiversity conservation. Established at the United Nations Climate Change Conference in Copenhagen in December 2009 and supported by the United Nations Food and Agriculture Organization (FAO), RTOACC participants spent their first year engaged in activities such as quantifying the climate benefits of organic farming which can be used for building up carbon-offset methodologies approved for international emission, and developing and improving life cycle assessment (LCA) tools for a better integration of organic farming techniques.

This is not to say that there is a dearth of knowledge on the role of organic agriculture in mitigating climate change. The fact that organic farmers replace synthetic fertilizers with biomass management results not only in enhanced soil fertility, but also increased soil carbon sequestration.

What RTOACC can contribute, through its multi-stakeholder platform, is setting base values so that all future investigation, methodology development and quantification work moves ahead from the same point. RTOACC participants use these base values as a point from which to identify what data is available and what data is missing, to identify current or develop new methodologies that can fill the data gaps, and then to use the new complete data to quantify the mitigation potential of organic agriculture. It is well known that there have been no relevant studies on soil carbon stocks in Africa or South America so further investigation will be required in order to access and incorporate reliable data from those areas. This information not only can enhance climate change mitigation activities which will have broad benefits, it also can provide the data to verify the mitigation benefits of organic agriculture which will allow organic farmers to increase their participation in carbon markets.

Looking to the future, RTOACC is committed to making a concerted effort to disseminate its findings to and through a variety of communication networks. For example, results will be sent to scientific publications to build a broad peer-reviewed knowledge stock that can be taken into account by the IPCC and other relevant scientific institutions; to national GHG inventories to develop management-specific information for their agricultural segments; and to data bases to share the knowledge of specific inputs and techniques of organic agriculture. In addition, RTOACC can share its newly improved or developed methodologies to appropriate entities to facilitate approval of organic practices for the regulated and non-regulated carbon markets.

Looking at the progress made in its first months of operation, RTOACC can look back at a time of fruitful activities and be proud of what its participants have achieved.

Urs Niggli
Research Institute of
Organic Agriculture

Nadia El-Hage Scialabba
Food and Agriculture
Organization of the United
Nations

1

ROUND TABLE ON ORGANIC AGRICULTURE AND CLIMATE CHANGE (RTOACC)

RTOACC meeting in Frick, Switzerland, May 2010

RTOACC MANDATE

Justification and membership

Considering that:

- climate change is one of the most serious challenges facing nations, governments, business and citizens over future decades,

- climate change directly influences food production and will act as a multiplier of existing threats to food security and malnutrition,
- small holders in rural areas will be most vulnerable to this change,
- agriculture is estimated to account for 10–12 per cent of global greenhouse gas (GHG) emissions and is responsible for 47 percent and 58 per cent of total anthropogenic emissions of Ch4 and N2O, respectively (Smith *et al.*, 2007),
- the recent "health Check of the CAP" of the EU has identified climate change as one of the new challenges for agriculture,
- organic agriculture aiming at producing food has a huge potential to mitigate climate change through soil carbon sequestration, reduced GHG emissions and sustainable use of natural resources,
- this mitigation potential is system immanent to organic farming as are efficient resource use (water, nutrients), food security and ecosystem services,
- climate change should influence consumer behaviour and climate-friendly practices must be adopted not only by farmers but also by food processors and retailers,
- there are still significant uncertainties around evaluation of GHG emissions and carbon sequestration related to natural processes,
- there is a potential for improvement within organic production and processing, organic food and organic supply chains,
- there is a comparative advantage in organic farming systems,

The following proponent organizations have agreed on the establishment of a Round Table on Organic Agriculture and Climate Change (RTOACC) in December 2009.

- Food and Agriculture Organization of the United Nations (FAO), Italy,
- Research Institute of Organic Agriculture (FIBL), Switzerland,
- International Centre for Research in Organic Food Systems (ICROFS), Denmark,
- Rodale Institute, USA,
- Environmental and Ethical Certification Institute (ICEA), Italy,
- Soil Association, UK,
- KRAV ek för, Sweden,
- Organic Federation of Australia (OFA), Australia,
- International Federation of Organic Agriculture Movements (IFOAM), Germany.

Since the establishment of the RTOACC, the following institutions have joined the proponent organizations:

- Agricultural and Processed Food Products Export Development Authority (APEDA), India,
- Andalusian Committee for Organic Agriculture' Association (CAAE), Spain,
- Bio Inspecta, Switzerland,
- Corporación Educativa para el Desarrollo Costarricense (CEDECO), Costa Rica,
- Danish Agricultural Advisory Service (DAAS), Denmark,
- Green Chemistry Bionet, Italy,
- Hivos, the Netherlands,
- Institute for Agriculture and Trade Policy (IATP), USA,
- Louis Bolk Institute, the Netherlands,
- Organic Research Centre, Elm Farm, UK,
- Soil and more International, the Netherlands,

- Spanish Society for Organic Farming/Agroecology (SEAE), Spain,
- Textile Exchange, USA.

The RTOACC is a multi-stakeholder initiative

The RTOACC implements transparent, fair and participatory governance. New members agree to support organic agriculture, the organic movement and its potential to address climate change, in line with what is mentioned in the considerations and aims above. members promote and communicate this commitment throughout their own organization. As per 'Criteria of Collaboration and Admission', RTOACC shall be composed of two categories of members: Participating members representing organizations and Observing members (individuals).

The RTOACC defines the process for admission and is open for new Participating members from the following sectors:

- associations in the organic sector, standard-setting organizations and certification bodies operating in organic agriculture,
- environmental organizations,
- intergovernmental organizations (IGOs),
- organization involved in the management of the voluntary offset mechanisms
- and the clean development mechanism (CDM),
- research organizations,
- funding organizations.

The RTOACC should aim to ensure the representation from the global south.

RTOACC aims to

- promote the potential of organic farming to mitigate climate change,
- promote the potential of organic farming as a climate change adaptation strategy,

- promote innovation, research, standard setting, and awareness building about the advantage of organic farming systems,
- initiate, support and facilitate the research on organic agriculture and climate change,
- identify viable ways of adaptation to the impacts of climate change,
- develop and implement services that support smallholders,
- advise the international community on organic agriculture and climate change issues, with a view to initiate policy change to wider adoption and support of organic agriculture,
- support the RTOACC member organizations as well as other governmental and non-governmental organizations in developing and fully implementing policies on climate change,
- advise in the development of climate-related provisions in international standards,
- initiate and support the development of a methodology to enable a reliable quantification and certification of GhG emissions and carbon sequestration at the various stages of the production process, and the identification of potential mitigation measures,
- support management practices and standard development issues that look at improving organic standards from a climate change perspective.

The RTOACC establishes:

Art. 1 – Initiate and facilitate research on organic agriculture and climate change

Therefore, the RTOACC will support:

- basic and applied research on assessment and dissemination of state of the art knowledge on the mitigation and adaptation potential of organic agriculture,

- compilation of consistent data for organic agriculture as a basis for the assessment of the climate change impact of organic agriculture,
- identification of the research gaps in this context, supporting and commissioning research to fill those.

Art. 2 – Adopt, further develop and disseminate concepts and methodological frameworks for measuring GHG mitigation and carbon sequestration in organic agriculture

The RTOACC adopts, further develops and disseminates concepts and methodological frameworks for measuring and accounting the GHG mitigation and carbon sequestration in organic agriculture worldwide. The positive effects of organic agriculture on climate change, once calculated through the adoption of proper tools, should be valorized within the existing offset voluntary or mandated initiatives. To do so, the RTOACC will identify the best partners to co-operate with.

Art. 3 – Provide information for improving awareness and technical knowledge on climate change

The RTOACC intends to develop information on climate change to be addressed to the organic farming sector, policy makers, farmers and food retailers. The aim is to increase understanding of the potential implications of climate change and of the available opportunities for adopting measures and best practices to address the double challenge of reducing GHG emissions while at the same time adapting to projected impacts of climate change.

Art. 4 – The RtOACC provides support for advocacy concerning the advantage of organic farming in adaptation and mitigation regarding climate change

The RTOACC intends to support the international community on organic agriculture and climate change issues, with a view to initiate policy change to wider adoption and support of organic agriculture.

Art. 5 – Explore possibilities to develop organic standards to a higher level of climate performance

The RTOACC intends to share results from its own and partner's work to the attention of concerned standard setters. The aim is to allow for organic standards to develop in a direction of better climate performance.

WORK Of RTOACC IN 2010

Climate change directly influences food production and food security. Throughout the world, dedicated agronomists, crop scientists and food producers across a myriad of national, regional and international organizations are working to increase understanding of the threat, and developing tools and methods for either slowing the progression of climate change or enabling food producers to adapt to changes that cannot be stopped.

RTOACC brings together stakeholders and partners from across the organic food production chain to discuss organic farming's potential to mitigate and adapt to climate change. Its principal objectives are to initiate, support and facilitate research on organic agriculture and climate change, advise the international community on organic agriculture and climate change issues, and develop a measurement method to enable reliable quantification and certification of carbon sequestration in organic agriculture. Its activities are supported by the Food and Agriculture Organization of the United Nations (FAO).

Throughout its first year, the Round Table's main activity was working to identify available data as well as to pinpoint data gaps that need to be filled in order to develop an organic agriculture methodology for the carbon market that would synergize with general development goals and also potentially benefit smallholders in poor countries. This publication presents the output of three Round Table workshops held to address these issues.

Workshop 1: "Data consensus and data gaps related to soil carbon sequestration potential of organic crop and livestock systems", and "Organic farming systems' potential for accreditation of a methodology for the carbon market", hosted by the Swiss Institute for Organic Farming Research (FiBL), 10-11 may 2010, in Frick, Switzerland.

> Participants discussed the mitigation potential of organic agriculture, related data availability and data gaps, the potential for organic agriculture in the carbon markets and strategies for the role of organic agriculture in climate policy. Initial workshop discussions focused on establishing common ground. This included an in-depth scientific exchange covering the available data on GHG emissions and the carbon sequestration potential of organic crop and livestock systems. As participants identified gaps in the available data, they shared ideas on how they could be filled. They also addressed the climate change mitigation potential of organic farming systems, the potential for developing methodologies for the carbon market, and the potential parallel benefit offered by carbon trading in terms of enhancing farmer income and food security. This information, presented in Chapter 2, enabled workshop participants to work on further developing the institutional context for harmonizing and improving the knowledge related to GHG emissions, carbon sequestration and carbon trading

systems in the RTOACC community. Based on their discussions, participants decided to strengthen and coordinate collaboration, information exchange and research through the RTOACC and further develop the expertise of the RTOACC and its members. They also committed to initiating outreach activities to increase the international community's awareness of organic agriculture and climate change issues.

Workshop 2: "Life cycle assessment of organic food and farming systems: focusing on greenhouse gas emissions, carbon sequestration and methodological challenges and status", organized by the International Centre for Research in Organic Food Systems (ICROFS) on behalf of RTOACC, 21 September 2010, in Bari, Italy. It was held in combination with the VII International Conference on Life Cycle Assessment in the Agrifood Sector.

The workshop focused on LCA methods, models and databases with a focus on GHG emissions and sequestration potential of organic food and farming systems. Participants heard invited speakers explain key methodological issues in LCA for organic systems and reports on carbon sequestration in organic agriculture and soil organic matter building in smallholder farms in tropical countries, respectively. The presentations and the results from the discussions at the workshop have provided important input to this report. Specifically, the overview of research studies comparing GHG emissions from organic agriculture with conventional agriculture, and the suggested methodological developments presented in Chapter 3 were partly developed for the meeting and reflect the discussions and comments given by the experts at the meeting.

Workshop 3: "Review of progress and follow-up", hosted by the Food and Agriculture Organization of the United Nations (FAO), 22 November 2010, in Rome, Italy.

The results of the work of the two previous workshops led the base for future work to be undertaken by the RTOACC, including related climate change science, climate labelling and awareness raising. Strengthening the scientific knowledge of organic agriculture and climate change has started by addressing mitigation issues, such as soil carbon sequestration and LCA approaches, but efforts are required to address adaptation issues, such as resilience brought by green chemistry. Such topics need to be investigated through integrative research approaches that combine field research with modelling and up-scaling tools for measurement, reporting and verification. The development of climate-relevant standards and labelling requires a review of existing standards relevant to organic agriculture. Raising awareness on RTOACC work includes communicating results to experts involved in the preparation of the Fifth Assessment Report of the Intergovernmental Panel on Climate Change and advocacy and partnerships with the international community concerned by sustainability and climate change[1].

As this publication shows in its reports from the workshops, the RTOACC members have found considerable mitigation potential in organic agriculture. however, supporting this mitigation potential in the context of existing mitigation policies and institutions poses considerable challenges. Quantification of the mitigation potential of organic in comparison to conventional agriculture still needs much research and a much broader data basis than currently available. Trends that support the beneficial performance of organic agriculture are visible in soil carbon sequestration in temperate climates, but consolidating these findings on a global level will require more data from the tropics and subtropics. This will be a main subject of ongoing work among the members of the RTOACC and with other institutions.

Linked to the difficulties of quantification, the offset mechanisms are accessible for single sustainable practices in

---

[1] RTOACC workshop presentations are available on the dedicated website: http://www.organicandclimate.org/ workshops.html

agriculture only, not for whole systems approaches such as organic agriculture. A conversion from conventional to organic agriculture as a mitigation measure would be too unspecific and heterogeneous to be accountable in established mitigation institutions. Focusing on single practices – such as composting and compost use, reduction of synthetic fertilizers, avoided biomass burning or agroforestry – is not a disadvantage, as supporting sustainable practices is beneficial in any context and fits ideally to organic systems. A further development of adequate institutions to capture the mitigation potential of organic agriculture for global mitigation policies will also be an important topic for the RTOACC members.

2

SOIL CARBON SEQUESTRATION OF ORGANIC CROP AND LIVESTOCK SYSTEMS AND POTENTIAL FOR ACCREDITATION BY CARBON MARKETS

Andreas Gattinger, Adrian Müller, Matthias Häni, Bernadette Oehen, Matthias Stolze and Urs Niggli[2]

©FiBL/T. Alföldi

[2] All authors work in the Research Institute of Organic Agriculture (FiBL) Frick, Switzerland, except Matthias Häni who is affiliated to the Swiss Federal Institute of Technology (ETH), Zurich, Switzerland.

During a RTOACC workshop hosted by the Research Institute of Organic Agriculture (FiBL) in May 2010, participants discussed the potential for organic agriculture in carbon markets and the need to develop strategies for the role of organic agriculture in climate policy. To move in this direction requires quantifying and raising recognition of the mitigation potential of organic agriculture. Thus, the participants also looked at available data and began a process of identifying data gaps. In doing so, they presented the related ongoing work of their organizations and drew conclusions for the further orientation and actions of the RTOACC. This Chapter synthesizes the discussions and outcomes of the workshop.

EXISTING GAPS IN DATA REQUIRED TO QUANTIFY THE MITIGATION POTENTIAL OF ORGANIC AGRICULTURE

Efforts to assess the mitigation potential of organic agriculture still face huge challenges and data gaps. In order to meet these challenges, it is first critical to set base values and then to combine model, experimental and real farm data to reduce the work load required for establishing factorial field experiments. This requires determining how sophisticated the data collection should be and the "type" of organic agriculture that will be included. RTOACC has identified the areas where data is most needed as:

- input-related emissions, such as from compost or fertilizer preparations,
- process-related emissions and emissions from various management types, such as legume rotations, reduced tillage, N_2O dynamics of compost application and soil carbon,
- emissions and soil carbon sequestration of entire production systems,
- emissions of specific crops within complex spatially diverse crop rotation systems.

In order to find ways to fill these data gaps, efforts are underway to set parameters and identify steps for ensuring consistency of data. For example, this could include standardizing key parameters such as: emissions factors for CH_4, N_2O, CO_2, soil carbon stocks and thickness of soil horizons, making use of existing long-term trial, establishing a database, defining standards for data quality and building up a body of knowledge.

Those committed to filling these data gaps face a trade-off between detailed and reliable data that require correspondingly expensive measurement approaches on the one hand, and fast, widely applicable and inexpensive measurement approaches that have correspondingly less detail and reliability on the other. Adequately identifying and supporting mitigation in organic agriculture requires finding a balance between scientific approaches based on detailed empirical data, and those based on broader visionary and conceptual approaches. This means determining which indicators and weights will assess the performance of a certain system against different indicators with respect to mitigation and co-benefits. For example, aggregation into a one-dimensional indicator can be avoided by using multi-dimensional spider diagrams to compare systems and inform decisions. however, it remains important to avoid focusing solely on organic agriculture as a mitigation instrument. It is also necessary to promote its other equally important benefits such as animal welfare, biodiversity, soil fertility and ethics.

RTOACC is committed to contributing to closing these data gaps and providing the scientific basis for decisions on balancing sustainability indicators. many RTOACC members have specific research underway that is producing relevant data on the mitigation potential of organic farming, such as a meta-study on soil carbon conducted at FiBL (presented further in the text), and two assessments of soil carbon sequestration: one under mediterranean site conditions conducted by the Spanish Society for Organic Farming (SEAE) and one under tropical site conditions conducted by Educative Cooperation for the Development of Costa Rica (CEDECO).

Organic agriculture can offer sustainable carbon credits. Although the financial rewards of the credits will likely be moderate, they could support financing the transition from a conventional to an organic system or the adoption of certain climate-friendly practices in both plant and animal production. In addition to their mitigation impact, credits related to organic farming practices offer a variety of valuable co-benefits, such as their indirect contribution to food security, yield stability, sustainability and adaptation to climate change, as can be seen specifically in plant and animal production:

- In plant production, the potential for generating carbon credits is mainly seen in compost use, biomass waste and manure storage and handling, fertilizer avoidance, biogas production, agroforestry and in avoided biomass burning. Due to the huge areas under agricultural production, soil carbon sequestration has a considerable global mitigation potential, although the potential per hectare is usually rather low and thus not ideal for the existing carbon crediting mechanisms.
- In animal production, the main potential for generating carbon credits is seen in improving lifetime performance by reducing GHG emissions per unit of output. The reduction of concentrate feed has a huge mitigation potential due to the land-use impact of concentrate feed production. however, capturing this in the existing carbon crediting mechanisms will be difficult, mainly due to the global system boundaries often involved. The potential co-benefits of these credits are manifold such as increased energy efficiency, improved livelihoods, improved biodiversity and soil organic matter, and longer term soil fertility, system stability and resilience.

Credit-based approaches to organic agriculture face specific challenges due to the rather low level of financial flows involved and the need for optimal institutional organization to manage payments from carbon finance. Assessing the carbon price necessary to make mitigation projects in organic agriculture attractive and relevant to farmers requires detailed data on farm economics. Furthermore, due to the low mitigation potential per hectare, several hundred to several thousand farms need to be grouped in order to be worthwhile. In such a context, the organic certification system may offer opportunities to simplify monitoring.

Application of certain techniques has potential to make organic agriculture more efficient; however, it will require coordinating a complex set of measurement methods and indicators for a complex set of different farm types. The workshop suggested the establishment of an organic agriculture-climate change board as a straightforward solution. Properly designed, it could plan and coordinate efficient management and application of all this knowledge, through a pragmatic learning-by-doing exercise.

©Marie Trydeman Knudsen

Organic banana and coffee field

At the same time, due attention must be given to the incentive and fairness aspects of carbon payments. At this point, only farms changing their management from conventional to organic can apply for these payments. This means that farmers who already converted to organic management, and thus already run their farms sustainably, do not receive such payments.

Looking at the long term, carbon finance institutions need to recognize that carbon credits and carbon trade do not provide the best solutions for supporting organic agriculture.

RTOACC suggests an approach based on voluntary agreements, using local markets that can build on trust, as opposed to global approaches based on high monitoring requirements. The design of more appropriate policy instruments is another option. These options would have better chances of adoption if, for example, they were based on the idea of combining taxes with subsidies or offered grandfathered emissions payment schemes.

Organic agriculture has the potential to play an important role on the more aggregate level of the newly emerging general approaches in climate policy, such as Nationally Appropriate mitigation Actions (NAMAS) or National Adaptation Programmes of Action (NAPAS). It also has to be emphasized that the performance of organic agriculture would be advantageous in even broader approaches to climate policy, based on the internalization of external costs, such as through national or global carbon taxes.

POTENTIAL OF SOIL CARBON SEQUESTRATION OF ORGANIC CROP AND LIVESTOCK SYSTEMS

This section presents a meta study on the carbon sequestration potential of organic agriculture, including the study aims, methodology and results.

Introduction

In 2010, FiBL conducted a literature review on soil organic carbon (SOC) contents, stocks and sequestration rates in organically managed soil, using 45 suitable scientific papers and 280 different data set, and undertook a quantitative evaluation of the obtained results using meta-analysis.

Meta-analysis, a statistical procedure that combines data from multiple studies, allows a quantitative proof of a hypothesis and offers a significant advantage over a narrative review that does not allow a quantitative proof of a given phenomenon. Although used mostly in medicine, for example to combine results of clinical studies, meta-analyses can be applied to other disciplines as well, and outcomes can be used to discuss and identify effective applications. In contrast to conventional statistical procedures, a meta-analysis takes the sample sizes and significance levels of single data sets into account when calculating the main effect size. This makes it an ideal tool for assessing an entire knowledge area, determining a reliable, average main effect size, and identifying research gaps.

This meta-study had two major goals:

- quantify SOC contents, stocks and sequestration rates in soils under organic and non-organic management,
- analyse factors influencing soil carbon levels.

The factors analyzed included climate, soil texture, land use (arable, grassland, horticulture), management type (organic or non-organic), crop rotation (with or without grass-clover leys), fertilizer type (with or without organic manure) and fertilization level (below or above 1.4 livestock units per hectare).

Material and methods

Only studies based on pair-wise comparisons (under similar site conditions) for organic and non-organic farming practices were considered. In one case, a fertilizer experiment was included (manure vs. mineral), but all other studies were based on farming system comparisons, where the organic practice was exclusively defined as "organic" by the authors.

Sugar cane field

This study followed five steps:

- literature search,
- literature review and evaluation,
- integration into data matrix and parameterization of those studies determined to be positive, meaning they contained a pair-wise comparison of organic vs. non-organic,
- descriptive and explorative statistics with the Statistical Package for the Social Sciences (SPSS) data mining software,
- meta-analysis with Comprehensive meta-Analysis (CMA) software.

Online information resources were searched for published studies, using the search terms "carbon AND soil AND conventional" in abstract, title and keywords. The resources searched included: CAB Abstracts, Google Scholar, ISI Web of Knowledge and Conference Proceedings, BIOSIS Previews, Scopus, SCIRUS, AGRICOLA, Scielo, GeoRef database, ScienceDirect and Organic Eprints.

Because of poor data sources from developing countries, recognized experts in organic agriculture, carbon, soil sciences or other relevant fields of research were contacted to contribute further ideas on resource identification and invited to share relevant publications or data. Furthermore a "Call for soil carbon data" was placed as a poster at the Tropentag International Conference on Research for Development in Agriculture and Forestry, Food and Natural Resource management in Zürich in September 2010 and the literature search remained open until this manuscript was submitted in spring of 2011.

Any publication assessed as positive for the approach was integrated into the data matrix and parameterized accordingly. Descriptive and explorative statistics were computed with SPSS software and meta-analysis with CMA software. The meta-analysis tool allows for a quantitative evaluation of published data taking observation points (= sample numbers) and variation of the target variable (i.e. Soil Organic Carbon (SOC)) into account.

Results

Descriptive statistics. In the initial stage, 45 publications were integrated into the data matrix: 37 peer-reviewed papers from scientific journals, and eight peer-reviewed conference proceedings, book chapters or dissertations. All 45 publications are based on pairwise system comparisons, from 44 field research projects consisting of 21 long-term plot experiments, five field trials and 18 farm comparisons. They encompass 280 data sets (lowest data aggregation level: general statistics) based on 2 477 samples (meta-analysis).

Explorative statistics. The average duration of management of all included studies was 16.7 years, with the oldest found in Europe, as shown in Figure 1. No relevant Africa or South America studies were found, so those continents are not represented in the study, as shown in Figure 2. The sampling depths of the different SOC studies varied between 8 and 60 cm, as shown in Figure 3. However, most of the samplings were performed down to 20 cm, with an average recorded soil depth of 22.5 cm. In this first analysis, the total sample number (N) was 2 477. A simple comparison of the data sets (N=280) by analysis of variance (ANOVA) revealed that organically managed soils contained higher SOC contents (concentrations as expressed in mass percents) than conventional soils (Figure 4). The same was true for the SOC stocks (i.e. absolute masses; N=118), even though fewer studies contained data of bulk densities which are necessary to calculate SOC stocks (Figure 5). In soils under organic management, the SOC stocks averaged 37.4 tonnes C ha^{-1}, in comparison to 26.7 tonnes C ha^{-1} under non-organic management.

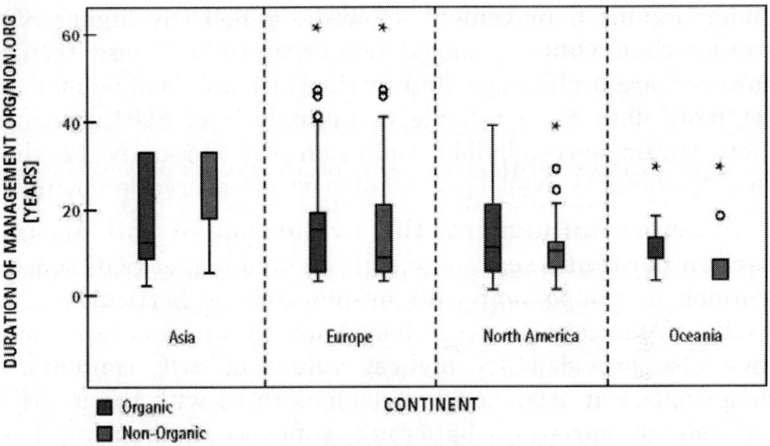

Figure 1

Monitoring length of different management practices (organic and conventional) considered in the farming system comparison (N=2 477)

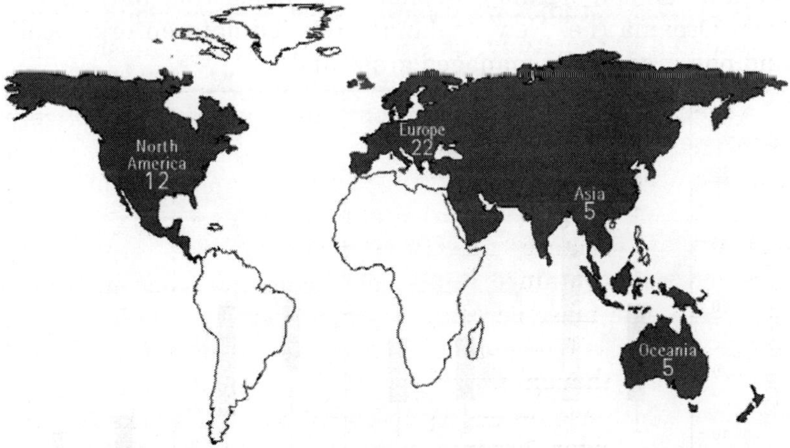

Figure 2

Geographical distribution of the sample of soil carbon studies used in the pair-wise comparisons of organic and non-organic management

Meta-analysis of soil organic carbon contents and stocks. The meta-analysis of SOC contents and stocks revealed the same result as had been determined by ANOVA and explorative statistics. meta-analysis revealed that soils under organic management showed significantly higher SOC **content** than those managed non-organically (N=2 477) and that soils under organic management showed significantly higher SOC **stocks** than those managed non-organically. These results, however, are preliminary and further attempts will be made to get more data for a reliable meta-analysis of SOC stocks as there are far fewer eligible studies on SOC stocks (N=12) than on SOC contents (N=2 477) and also fewer observation points.

Factors influencing the evaluation of soil organic carbon contents. Grassland soils showed higher SOC concentrations in comparison with arable land or horticulture. As studies from Oceania were based mostly on grassland data, they also provided the highest values of SOC contents. A somewhat clear tendency was demonstrated with the multiple analysis of variance that ranked factors influencing SOC contents. The analysis found that climate had the strongest impact on soil organic carbon contents followed by land use (arable, grassland, horticulture) and the management system (organic or non-organic). It should be noted that only studies from Oceania (i.e. New Zealand) provided data on organically and non-organically managed grassland.

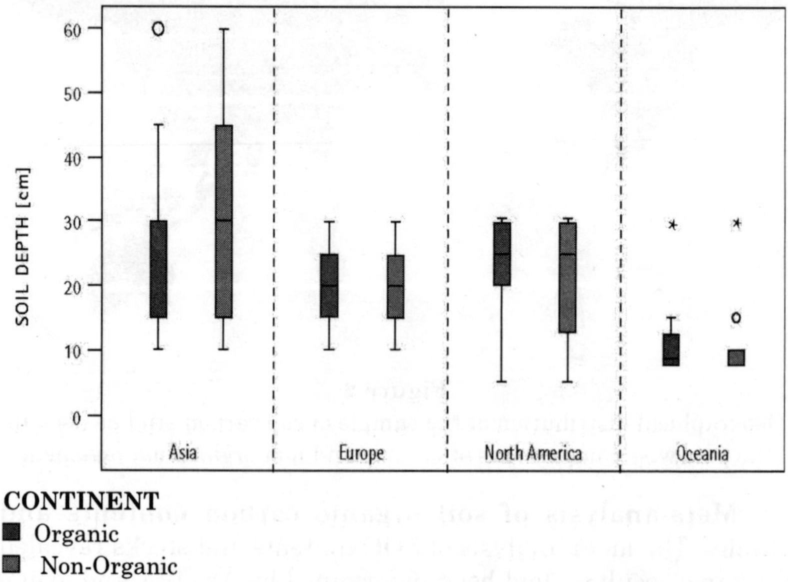

Figure 3

Variation in sampling soil depth of different analyzed soil carbon studies (N=2 477)

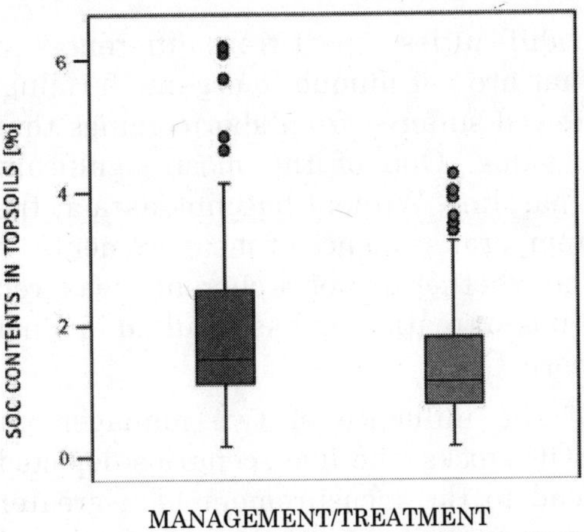

mean	N	Std. Deviation
1.630145	280	1.0946869

Figure 4

Soil organic carbon (SOC) contents (expressed in %) are significantly higher in organically managed soils

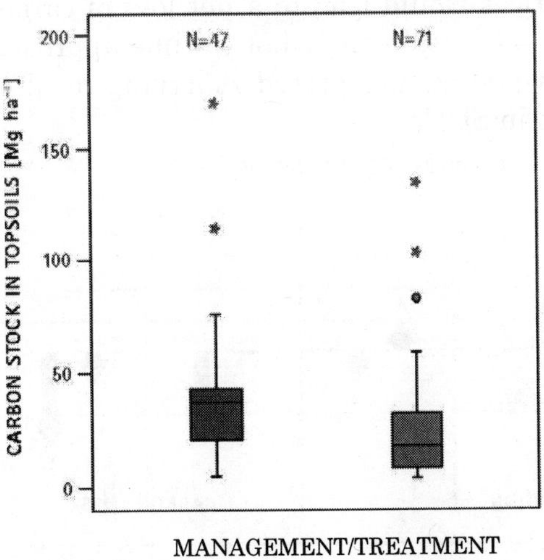

mean	N	Std. Deviation
30.081845	118	26.4863352

Figure 5

Soil organic carbon (SOC) stocks (expressed in tonne of carbon ha^{-1}) are significantly higher in organically managed soils

Methodological difficulties of the meta-study

The baseline problem. Efforts to determine soil carbon sequestration in organically managed soils face manifold data gaps and methodological difficulties. Apart from differences in management practices that are not unique to organic farming, many of the studies reviewed suffered from shortcomings that reduced their scientific value. One of the most significant limitations was with the baseline. Without baseline data at the inception of a trial or a temporal sequence of measurements, it is impossible to determine whether or not a current measured difference in SOC between two treatments has resulted in a net sequestration of atmospheric CO_2.

In a comparison of the influence of two management practices (A and B) on SOC stocks, the five scenarios depicted in Figure 6 would all lead to the measurement of a greater stock of SOC under practice A. however, a net sequestration of atmospheric CO_2 would only occur in three of the five scenarios (i.e. Scenarios 1, 2 and 3, Figure 6). In Scenario 1, both management practices would lead to a net sequestration, while in Scenario 5, both practices would lead to a net loss of carbon back to the atmosphere. yet, with a snapshot-in-time approach, both Scenarios 1 and 5 would be interpreted as having resulted in the same relative gain in SOC.

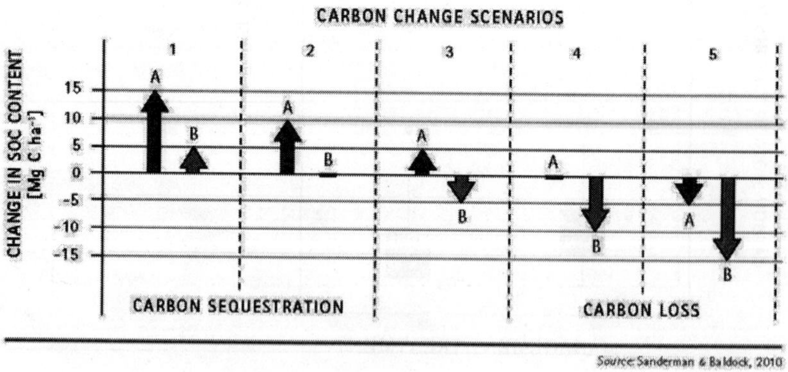

Figure 6
Five different scenarios of carbon change induced by two management treatments (A and B) after a set amount of time.
The arrows indicate the direction of carbon change and their size reflects the magnitude of carbon change. All five scenarios give the same relative difference of 10 Mg C ha^{-1} in SOC between treatment A and treatment B

A second consideration involved in defining the influence of applied management practices on SOC stocks was whether SOC has stabilized at a new steady state value indicative of the original management practice or is still changing and progressing towards a new equilibrium value. This consideration is often the underlying reason for the various scenarios in Fig. 6.

Evidence suggests that imposing agriculture on previously undisturbed soil will result in a 20–50 percent loss of SOC (Lal, 2004), with the rate of loss being greatest initially and then diminishing over time (dashed line in Figure 7) with a new equilibrium not reached for 20–100 years. In addition, different SOC sequestration outcomes will be obtained if two management practices (conventional and best practice in terms of SOC accumulation) are initiated at different times after clearing (points A, B and C in Figure 7).

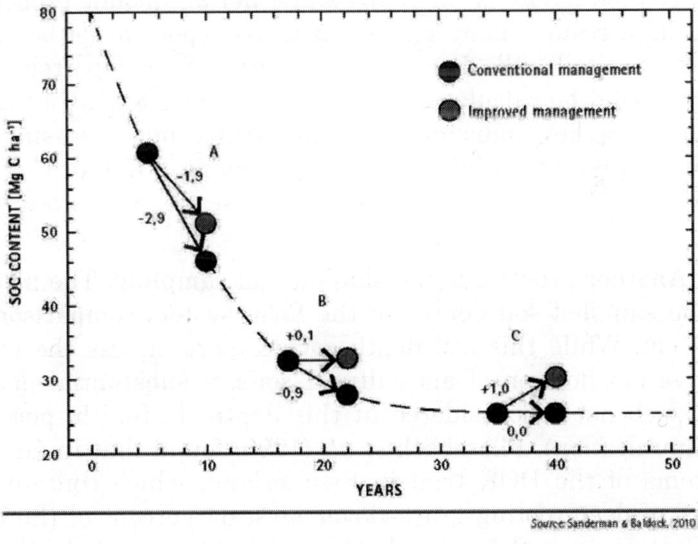

Figure 7

Hypothetical field trial simulation comparing conventional and improved management practices initiated at three different times (A, B and C) after converting a natural ecosystem to agricultural production in year zero

All three points show the same relative gain of 5 Mg C ha^{-1} in the improved management practice over a five year period; however, the actual rate of change is completely different

The relative difference in SOC content measured between the two management treatments at all three times is similar (5 tonnes of Carbon ha^{-1} over a five-year period). However, the benefit is completely different in terms of sequestration of atmospheric CO_2 relative to the conditions present at the start of the three experiments. Without SOC measurements taken at the start of each of the experiments (A, B and C), the different carbon sequestration scenarios depicted in Figure 7 would not be evident and the best management system may be inappropriately considered to have sequestered atmospheric carbon.

Missing bulk densities and shallow sampling. The majority of publications, identified above in the Results section, reported SOC concentrations rather than stocks. The great majority of these studies were originally designed to define the influence of agricultural management practices on plant dry matter production, grain yields and other agronomic properties and, as a result, many long-term trials reported neither SOC stocks nor soil bulk density. If the latter were reported, SOC stocks could be calculated as shown in Figure 8. SOC concentration is a key indicator for soil fertility but assessing the sequestration potential requires the amount of CO_2 or C stored in a given soil, which is the absolute mass, i.e. SOC stock = t C ha^{-1}.

Another problem is the shallow soil sampling. The median of the sampled soil depths of the farm system comparisons is 22.5 cm. While this soil depth covers more or less the entire cultivation horizon of agricultural soils, a substantial part of SOC will not be considered at this depth (P. Smith, personal communication). Fliessbach *et al.* (1999) found that in farming systems of the DOK trial in Switzerland, which contain two years of deep-rooting grass-clover leys, 64 percent of the total SOC stocks are deposited between 20–80 cm soil depths. In many parts of the world, organic farming systems are relying on the soil fertility build-up of deep-rooting grass-legume mixtures and on the incorporation of plant residues by deep-digging earthworms, making it quite likely that the currently available data sets underestimate the SOC stocks in organically managed soils. This is particularly significant considering that in deeper soil horizons, SOC seems to be more stabilized.

Radiocarbon analyses of microbial short-chain Phospholipid Fatty Acids (PLFA) from different soil depths showed that the PLFAs in surface soils were derived largely from fresh plant residues whereas the radiocarbon values of PLFAs at 30–45 cm soil depth suggest the contribution of more stabilized soil organic matter (Rethemeyer *et al.*, 2005).

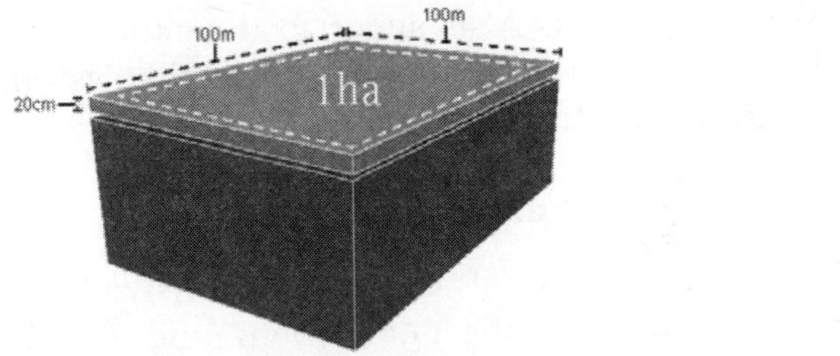

Carbon stock (t/ha) = soil organic carbon (g/kg) • bulk density (g/cm³) • depth (cm) SOC (g/kg) • BD (g/cm³) • depth (cm)

Source: Häni, 2010

Figure 8
Scheme and equation used to calculate soil organic carbon (SOC) stock

©Gustavo Fonseca de Almeida

Small-scale organic farmer

Poor data availability for major cropping systems and continents. In addition to the fact that no peer-reviewed study containing farming system comparison and reporting SOC values exists for the African continent or for Central and

South America, the Asian continent is largely under-represented with only five studies (see Figure 2). Grassland, as a land use, is only covered by two studies from New Zealand, which does not reflect the reality at all. Grassland is the dominating agricultural land use in many parts of Africa and Central Asia (e.g. Mongolia), and pastoralism – as a traditional and sustainable land use system built on grassland farming – is not represented at all. Also major food commodities such as rice and many tubers are not reflected in the system comparisons found in the literature search.

Summary and conclusions

The core work of the comprehensive literature review integrated 45 scientific publications and 280 datasets into a meaningful data matrix. Quantitative evaluation of this comprehensive data set revealed strong scientific evidence for higher SOC contents in soils under organic farming. This is also in accordance with the findings of Leifeld and Fuhrer (2010); their evaluation of 32 peer-reviewed papers and 68 data sets revealed that after conversion, SOC contents in organic systems increased annually by 2.2 percent on average, whereas in conventional systems, SOC did not change significantly. There is a lack of SOC data for developing countries, with no farm system comparison data from Africa and Latin America, and only limited data on SOC stocks which is crucial for determining carbon storage in soil. While this means that C sequestration rates for organic farming practices cannot be assessed reliably at the moment, further attempts will be made to access more reliable data on soil carbon stocks.

Next steps

The data matrix for the meta-analysis on SOC in organic and non-organic farming systems will be further refined in a manner that allows the variables of soil texture (i.e. clay content), crop rotation, fertilizer type and fertilization level to be used for further statistical evaluation and for a scientifically sound assessment of the factors influencing SOC in agricultural soils at a global scale. Also, the authors of the publications considered will be asked for data on soil bulk density. meanwhile, the FiBL worldwide network is contacting more people, including those from developing countries, seeking further relevant data that will enable a sound meta-analysis on SOC stocks and a sound calculation of C sequestration rates. Further, FiBL will continue to conduct a literature search relevant to the SOC study.

The research topic "C sequestration in organically managed soils" is far from full exploration. Even with a scientific paper produced on the above-mentioned meta-analysis findings, some important land-use types, such as grasslands and agroforestry in Africa, have not yet been investigated on SOC in a pairwise system comparison. It is unrealistic to expect representative SOC data for major cropping systems from Africa, Asia and South America within a short time-frame. This means that further research will be needed to fill these data and knowledge gaps. In this regard, the RTOACC can serve as a platform to exchange ideas and promote the bilateral or multilateral research on C sequestration as influenced by organic farming systems. however, in future SOC investigations, the above described data gaps and methodological uncertainties should be taken into account.

POTENTIAL FOR ACCREDITATION OF AN ORGANIC FARMING SYSTEM METHODOLOGY FOR THE CARBON MARKET

This section considers existing and foreseen methodologies that will help quantify and simplify the understanding of the potential role of organic agriculture in the carbon market.

The methodology development aims to capture the mitigation potential of organic agriculture projects in developing countries for the carbon market. Of course, organic agriculture provides a range of benefits other than its mitigation potential. Its potential to provide carbon offsets, as well as many additional sustainability benefits, would translate into higher financial rewards for the farmers.

Carbon market context

There are only a few projects that deal with land use, land use change, forestry and agriculture in the Clean Development mechanism (CDM) (Figure 9). The forestry sector has a much higher share of the Voluntary Carbon market (VCM) than agricultural activities (Figure 10). Further developments will see an increase in forestry offsets, e.g. under the UN Collaborative Programme on Reducing Emissions from Deforestation and Forest Degradation in Developing Countries (REDD and REDD+). however, agriculture will increasingly gain importance, as reflected in the recent submission of methodologies and protocols aimed at capturing the mitigation potential of agriculture, mainly soil carbon and nitrous oxide via optimized fertilizer management, such as the World Bank Voluntary Carbon Standard (VCS) methodology for Sustainable Agricultural Land management (SALM) or the International Fertilizer Association "4R: right source, rate, time, place" approach applied in the new nitrous oxide emission reductions strategies from Canada and the USA (Government of Alberta, 2010; International Fertilizer Industry Association, 2009; Voluntary Carbon Standard, 2010).

Compared to 2008, 2009 saw several striking shifts in transaction volumes by project type. hydro projects experienced the most significant market share losses, dropping from 32 percent to 7 percent (16.4 to 3.2 mtCO_2eq); wind, from 15 percent to 8 percent of the market (7.7 to 3.4 mtCO_2eq); and energy efficiency, from 4 percent to 1.4 percent (2.1 to 0.6 mtCO_2eq). The reasons for agriculture's - and to a less extent forestry's - low share of the Voluntary Carbon market (VCM) are manifold. however, all are related to the complex biological systems involved, which are not standardized or as easily quantifiable as industrial processes. Thus, monitoring Reporting and Verification (MRV) is highly demanding for agricultural and forestry systems, as the relevant data is highly variable and default values are not reliably capturing a single project at hand. Project, which are unviable under the CDM are somewhat more viable under VCM where requirements can be considerably lower.

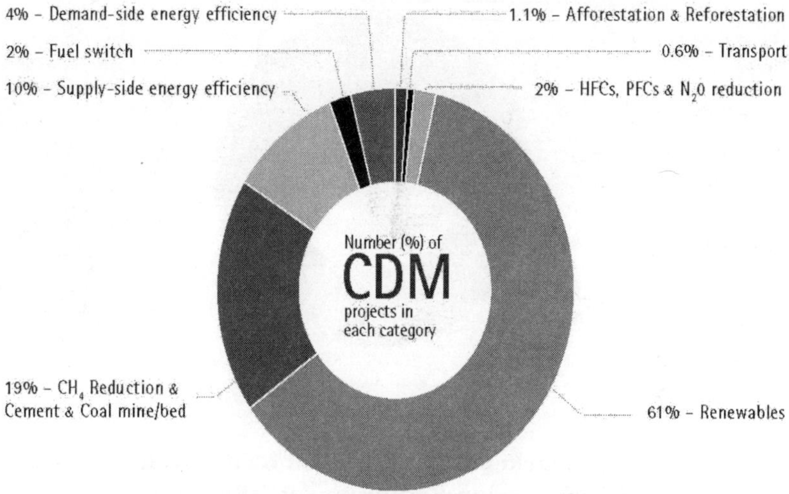

Figure 9

Certified Emissions Reduction (CERs) expected until 2012 from a number of different projects carried out in different sectors under the Clean Development Mechanism (CDM)

(a) Number and distribution in different sectors of existing projects (N = 5 600) as recorded in the CDM project pipeline, November 2010.

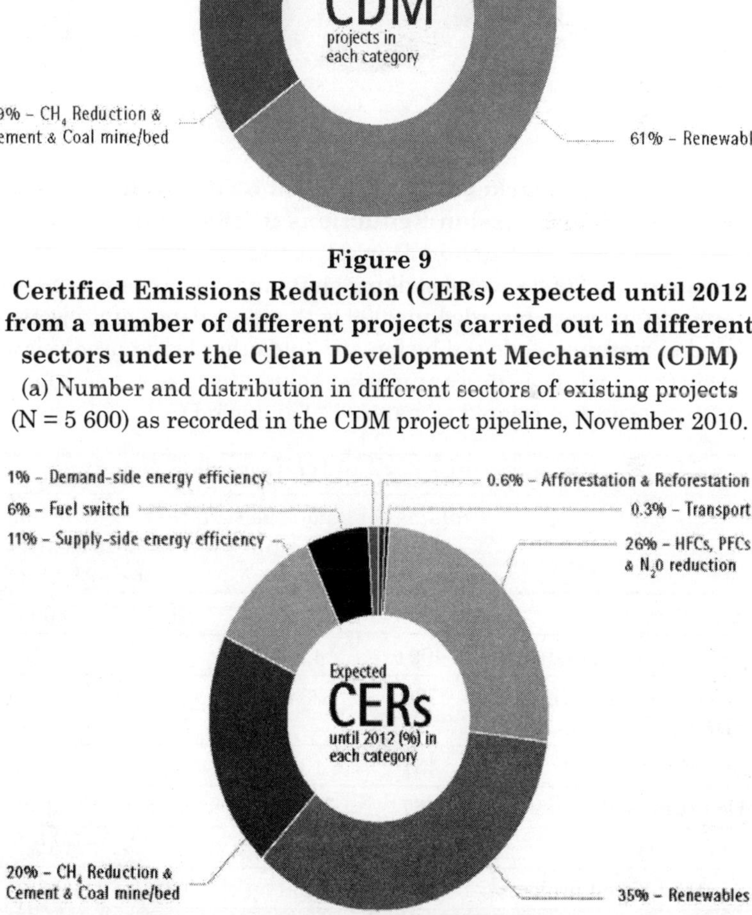

(b) Contribution of projects belonging to different sectors to the total certified GHG emission reduction in 2012 (Total CERs in 2012 = 2 800 Mt CO_2eq / 210 Mt CO_2eq traded in 2009).

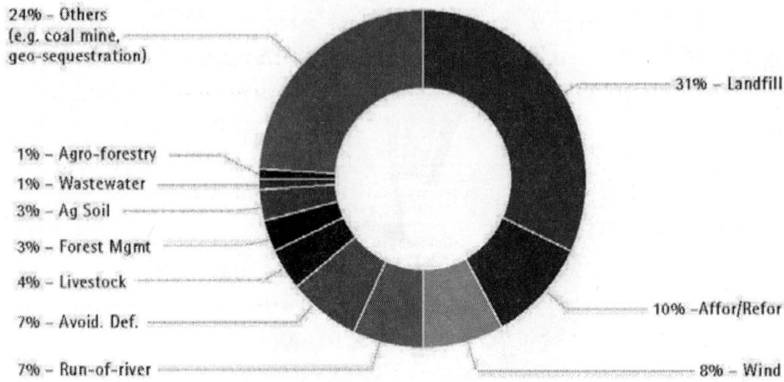

Figure 10
Percentage of market share achieved by different project types for Carbon Emission Reductions (CERs) in the voluntary Carbon Market (VCM)

(a) Compared to 2008 several striking shifts in transaction volumes by project type were recorded in 2009 with a prevalence of projects related to methane, followed by forestry and other land based related projects (24%) while significant market share losses were recorded for project related to water (7%) and wind (8%).

LAND-BASED CREDITS SOLD OTC, 2008 vs. 2009				
	Volumes of land-based credits ($ktCO_2eq$)		Market share of land-based credits relative to the total	
Project Type	2008	2009	2008	2009
Afforestation/reforestation	4091	4253	8%	10%
Avoided deforestation (REDD)	730	2846	1%	7%
Forest Management	431	1349	1%	3%
Agricultural soil	267	1250	0.5%	3%
Agro-forestry	-	625	-	1%
Other land-based projects	130	109	0.3%	0.3%
TOTAL	5650	10432	11%	24%

(b) The Carbon Emission Reductions (CERs) achieved by forestry and other land-based related projects passed from a market share of 11% (5.7 $MtCO_2eq$) in 2008 to a 24% (10.4 $MtCO_2eq$) in 2009.

A second barrier for land-based projects is the impermanence of generated credits, as they are mostly based on reversible land use change or management practices. In addition, due to the specific dynamics of the systems involved (soil, biomass growth, biomass waste, decay, etc.), issuance time can be considerably delayed in relation to project start.

Finally, profitability of such projects tends to be low, as they generate low numbers of credits per hectare. Thus, huge areas need to be covered, which again adds to the MRV problems. Forestry or agroforestry projects that have a higher density of credit generation per hectare are somewhat exceptions to this. Similarly, biogas projects and composting are more profitable, as their reliance on industrial processes in centralized plants reduces MRV costs. The MRV problems encountered in land-based projects were most recently illustrated, for example, by the rejection of the improved rice-cropping methodology NM0046 submitted to the CDM, which is largely due to a lack of knowledge on the underlying processes and their quantification or MRV[3].

Material and methods

The methodology development was based on an expert assessment of the current status of agriculture- related methodologies in the CDM and for the VCM. FiBL expertise on organic agriculture was combined with South Pole Carbon Asset management Ltd (SPC) expertise on carbon markets and the institutions of carbon finance. The assessment included the mitigation potential of organic agriculture and its wider sustainability performance when applied in smallholder contexts of developing countries, as well as the specific aspects of existing methodologies such as composting, optimized fertilizer use, N_2O protocols (in North America), rice production and agroforestry. This latter assessment was based on the original documents, expert comments from the stakeholder consultations on each methodology found on Internet, input from South Pole Carbon Asset Management Ltd (SPC), and personal informa-

[3] However, an improved rice methodology is now accepted.

tion from experts who participated in the RTOACC Workshop, and others. A particular focus was given on the reliability and viability of quantification, such as the MRV of the mitigation potential claimed on project level.

Results

A methodology for converting farming practices from conventional to organic management has no chance of being approved, as it is not specific enough. Thus, the approach focused on key practices in organic agriculture which can be captured in such a way as to make quantification of their mitigation potential compatible with the requirements from project-based offset mechanisms. For this, the aim was to develop a CDM methodology, as this is the most demanding and most respected standard. Knowing how a certain practice will have to be treated under the CDM, it can easily be simplified to meet lower standards, such as for the VCM.

Organic practices and characteristics of principal potential for carbon credit generation include:

- replacement of chemical fertilizers,
- production and application of compost,
- application of legumes in crop rotations,
- avoidance of burning agricultural waste and residues,
- increase of soil organic matter (e.g. soil carbon sequestration).
- however, the latter practice, soil carbon sequestration, is not as effective from

the carbon offset perspective as originally assumed, particularly when compared with mitigation practices involving methane emissions, such as optimized manure management, or methane recovery and biogas use from manure (see Figure 11). hence the decision was made not to develop soil carbon sequestration to a carbon offset methodology, at least initially.

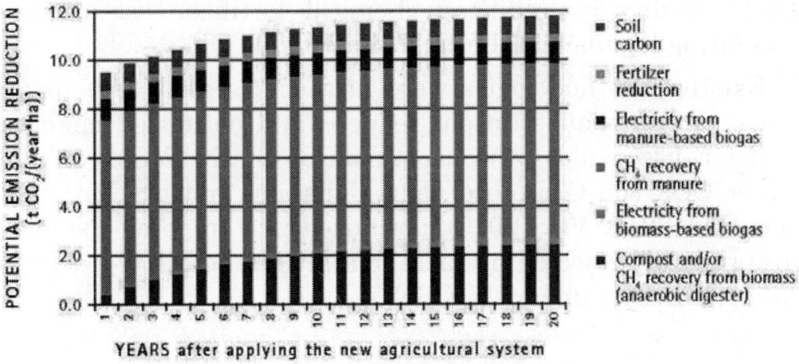

Figure 11

Rough and preliminary estimates of the potential of emission reductions achieved with mitigation practices applicable within organic agriculture

Source: based on calculations from South Pole Carbon Asset Management Ltd.

Further practices of importance to carbon capture include: agroforestry, restoration and less intensive use of peatlands, replacement of peat with compost in planting substrates, optimized rice production and certain processing steps such as those in wine and cheese making.

The decision was made to start with the "low-hanging fruit", regarding both the complexity of MRV and profitability regarding the number of credits per hectare. With the goal of capturing core practices of organic agriculture and the existence of methodologies for certain of the practices mentioned above (e.g. methane capture and biogas production, agroforestry), it was decided to revise the existing CDM compost production methodology (abbreviated as AMS.III-F) by adding biomass burning to the baseline, and mulching and optimal manure management to the project activity. In the same line, the AMS.III-R methodology was revised, which can be understood as a version of AMS.III-F specifically adapted to the context of smallholders through, for example, simplified MRV requirements. In order to capture the mitigation potential of organic agriculture regarding fertilizer application and soil carbon sequestration, a new methodology was developed, based on the existing CDM methodology AMS.III-A, which generates carbon credits by reducing chemical fertilizer use through inoculating legumes in the crop rotations.

Existing methodologies were further assessed with regard to soil-carbon sequestration, reduced and optimized chemical fertilizer use under various standards of the VCM (SALM, former CCX soil-C protocol, and Canadian and US N_2O protocols GoA 2010, VCS 2010), optimized rice cropping (Nm0046 was originally rejected by CDM, mainly due to MRV problems) and agroforestry under the CDM.

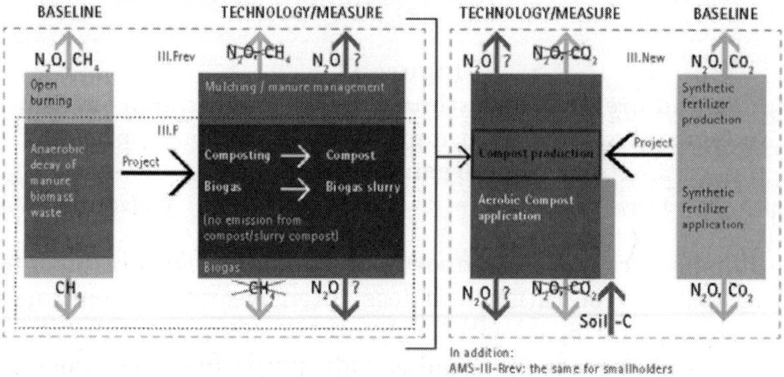

Figure 12

The interplay of the revisions of existing CDM methodology AMS.III-F and the new methodology based on AMS.III-A

Summary and conclusions

The tangible results of this work include two methodology revisions and a newly developed methodology for the CDM, which are now ready for implementation in existing projects. They capture organic waste, fertilizer management and soil dynamics (nitrous oxide/soil carbon) on farms in a consistent way, which is adequate for the particularities of organic farms and which can be captured in the institutional framework of carbon offset methodologies (Figure 12).

Adding biomass burning to the baseline in these methodologies is the most important revision. Biomass burning is a widely used and very unsustainable practice that has many adverse effects other than GHG emissions. It affects local air quality and leads to considerable nutrient losses. making avoidance of this applicable for the carbon market is an important step and generates sustainable carbon credits. Avoidance of biomass burning can be applied in a smallholder context, but it also makes sense on large scale, such as sugar cane plantations where pre-harvest burning is often the common practice. Furthermore, the avoidance of synthetic fertilizers and increased use of compost or mulching also improve resource and nutrient management. The revision of AMS.III-R also makes these opportunities available specifically to smallholders.

This work on carbon offset methodologies provided insights into the specific challenges that organic agriculture (and agriculture in general) faces when combined with the established institutions of carbon markets and offset mechanisms. Particular challenges are related to scientifically credible MRV (e.g. based on on-site measurements) *vs.* the practical applicability of MRV in a concrete project without incurring prohibitive costs (e.g. making heavy use of global default values). Other challenges relate to the comparability of outputs in the baseline and under the project activity. If crop rotations change, for example, the same-level-of-services assumption, which is important to avoid leakage of emissions, is difficult to assess and ensure. It remains open as to whether such assessments should be based on some monetarization or on other aggregation approaches, such as via energy contents. One solution to this problem currently adopted in certain CDM methodologies is prescribing crop rotations for the whole project lifetime and restricting phosphorus (P) and potassium (K) inputs under the project activity to the same levels as before. These conditions are clearly unviable, which likely is the reason for projects not to use the AMS.III-A methodology, which has these applicability conditions. Finally, profitability, and in relation to that, additionality of projects in agriculture remains a topic of concern, as the amounts of credits generated will remain relatively low. Assuring additionality will be less a problem when based on institutional rather than financial barriers.

Insights also were gained on ways to further develop carbon market institutions in order to account adequately for the specific characteristics of agriculture. For example, one approach called for refraining from undertaking project-based reduction in agriculture and instead capturing its mitigation potential in national strategies, based on a large number of projects where the default values for mitigation potential apply on average.

Next steps

FiBL will apply the two revised methodologies to existing projects in order to gain insights on their strengths and weaknesses in realistic settings. Subsequently, the methodologies will be further adapted and refined, in particular to include a monitoring section. Also, FiBL will prepare a Project Design Document necessary to submit the methodology revisions to the UN Framework Convention on Climate Change (UNFCCC).

However, for the time being, the new methodology on fertilizer application and soil carbon sequestration will not be applied, due to large scientific uncertainties. MRV requirements will either become prohibitively expensive or will remain scientifically weak, thus not leading to reliable mitigation accounting. It is however suggested to undertake revisions for the existing and submitted methodologies and protocols that contain fertilizer application and soil-carbon in order to make them applicable for organic agriculture as well, if possible. This will work for the World Bank VCS methodology SALM, but likely not for the Canadian N_2O protocol. Future data availability on soil carbon will also be monitored intensively. Given that the uncertainties and challenges of MRV can be reduced considerably, the new methodology on fertilizer application and soil carbon sequestration will be adapted and submitted to the UNFCCC.

As previously discussed, capturing the mitigation potential in agriculture on an aggregate level, such as in the context of NAMAs, seems more appropriate than capturing it via the established offset mechanisms. Project-based offsets in agriculture have a fundamental problem, due to the high variability of the biogeochemical processes involved and the correspondingly high uncertainty of emissions or mitigation in specific, concrete cases. Carbon offsets make sense in a context of standardized and reliably quantifiable processes, such as for emissions from industrial processes or energy generation. Beyond recognizing that the mitigation potential of single projects in agriculture cannot be quantified correctly, it is questionable how reliable it is to offset standardized and quantified emissions in industrial countries with emission reductions from highly uncertain agriculture mitigation in developing countries. On the other hand, on aggregate for the average of thousands of projects, the mitigation potential can be quantified, if reliable default values are available.

3

LIFE CYCLE ASSESSMENT OF ORGANIC FOOD AND FARMING SYSTEMS: METHODOLOGICAL CHALLENGES RELATED TO GREENHOUSE GAS EMISSIONS AND CARBON SEQUESTRATION

Marie Trydeman Knudsen[4], John E. Hermansen[4], Niels Halberg[5], Lise Andreasen[5] and Adrian Williams[6]

©FiBL/ A. Fliessbach

This chapter draws upon the discussions and recommendations from the Life Cycle Assessment of Organic food and farming systems workshop convened by ICROFS on behalf of

---

[4] Department of Agroecology, Aarhus University, Denmark

[5] International Centre for Research in Organic Food Systems (ICROFS), Denmark

[6] Cranfield University, United Kingdom

RTOACC in September 2010 in Bari, Italy. The workshop presented current knowledge on LCA methods, models and databases – specifically those that focus on the GHG emissions and sequestration potential of organic farming systems and organic food. It also discussed the improvements needed for LCA to be used by organic sector organizations and operators for the development of certification of energy and carbon labels and for improvement of the organic sector's climate impact. Methodological developments reflect the discussions and comments given by the experts at the RTOACC workshop. It should be kept in mind that this chapter focuses on the single category of climate change impact and does not cover all the important aspects of sustainability in relation to organic agriculture.

INTRODUCTION

Agriculture is responsible for 13 percent of global GHG emissions, while in Europe, food production and consumption are responsible for approximately one-fourth of the total GHG emissions (Foster *et al.*, 2006). This makes agriculture and food systems major players that must be considered in attempts to mitigate global climate change. If emissions from deforestation due to land conversion to agriculture were also taken into account, it could bring the total share of emissions from agriculture and food systems to approximately one-third of global anthropogenic GHG emissions (Scialabba and Müller-Lindenlauf, 2010). In the search for climate change mitigation options, the choice of production system and food supply and consumption might play a major role.

Organic agriculture offers an alternative production system and, to a certain degree, has a different food supply and consumption pattern (FDB, 2010). however, climate change mitigation has not traditionally been considered a benefit of organic agriculture. It is covered to a minor degree because it restricts use of mineral fertilizer which promotes the use of manure and, in turn, tends to increase the carbon sequestration in the soil. however, in order to comply with the organic principles (International Federation of Organic Agriculture Movements, 2005) and preserve credibility as the most environmentally friendly production system with regard to climate change, the effect of organic agriculture and food systems on GHG emissions needs to be investigated (Niggli *et al.,* 2008; Scialabba and Müller-Lindenlauf, 2010). While organic agriculture has a number of aims and possible benefits, such as those related to biodiversity, soil quality, animal welfare, avoidance of pesticides and fairness in the production chain, there is also a need to address how organic agriculture performs in terms of GHG emission, how this is assessed, and what options should be considered for improvement.

LCA is an internationally accepted method, recognized as the best tool to assess environmental impacts along the life cycle of a product (Finnveden *et al.*, 2009). At present, it is not designed to include the other environmental dimensions relevant to organic farming, such as climate mitigation through soil carbon sequestration and conserving biodiversity. however, it does include several environmental impacts and focuses on the product chain, which avoids shifting from one life cycle phase to the other or from one environmental problem to another (Finnveden *et al.,* 2009). For organic agriculture, environmental impact categories other than climate change are important, which makes LCA a valuable tool for organic agriculture and minimizes the risk of oversimplification. The methodology for identifying the carbon footprint of products with a single focus on GHG emissions is also based on LCA (Publicly Available Specification 2050, 2008).

The LCA methodology also has shortcomings, some of which are specifically visible when applying LCA to organic agriculture. For example, it has limited ability: to allocate and account for interactions in the farming system such as crop rotation effects, to estimate the environmental load from imported manure and to estimate and account for changes in soil organic matter (soil carbon sequestration) and land use. moreover, the sector is relatively small, which presents the challenge of securing representative data.

LIFE CYCLE ASSESSMENT METHODOLOGY

The LCA methodology is a tool to assess environmental impacts throughout the product chain. Recently, attempts have been made to include social or economic aspects in LCA, but traditionally the main focus has been on environmental aspects. The basis of LCA is a study of energy and mass flows.

The simplified basic data include:

- all inputs – including materials and activities,
- all production processes – including those upstream from the farm and those
- leading to emissions to the environment,
- all outputs – both wanted and unwanted.

These data enable the quantification of the flows of mass and energy within and across system boundaries. The ideal would be studying the products in a cradle-to-grave perspective, but often a field to farmgate perspective is applied for pragmatic reasons. LCA is a valuable tool for assessing GHG emissions related to a product, but it is also a very data-intensive and time-consuming tool. When using LCA for agricultural products, there are other challenges, especially with regard to land use, that need to be addressed, such as aspects of biodiversity, soil and land use change (Finnveden *et al.*, 2009).

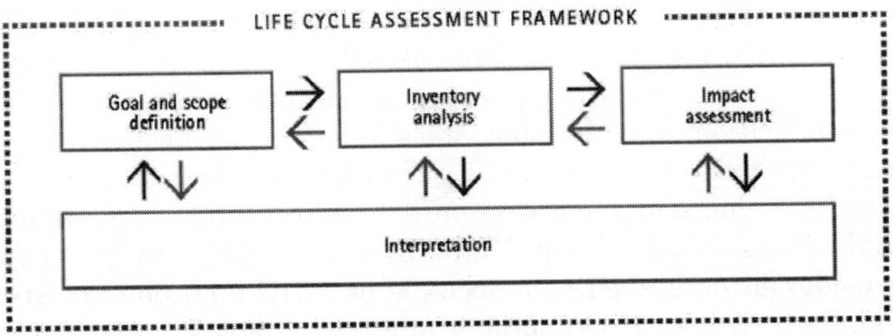

Figure 13

The four phases in the LCA methodology

When performing an LCA, the work is divided into four main phases, as illustrated in Figure 13. Each phase implies several methodological choices that may affect the final results differently, even when the ISO guidelines are followed (Finnveden *et al.*, 2009). The following sections describe the main aim and the major methodological choices in each of the LCA phases. Organic wheat is used as a hypothetical example in this description. The following looks at the phases of the LCA, including scope definition, inventory analysis, impact assessment and interpretation.

Goal and scope definition

LCA begins by defining the purpose, goal and scope of the study.

Purpose definition. It requires defining the so-called "functional unit", which is an exact description of the studied object, product or service.

Functional unit. This is related to the studied object, be it a product or service. As the LCA is a product-oriented methodology focusing on the product chain, the environmental impacts are expressed per functional unit, which could be mass, energy or protein (Roy *et al.*, 2009). In addition to the product chain, the LCA methodology also can be used in some cases to assess the environmental impacts per unit of area. The question of whether the environmental impacts should be expressed per product unit or unit area has given rise to debate (van der Werf *et al.*, 2007; Haas *et al.*, 2000; de Koeijer *et al.*, 2002). Some aspects are more related to the local or regional levels (such as land use in terms of biodiversity, soil and eutrophication) and therefore more relevant per hectare, while other are more focused on the global level (such as energy use and GhG emissions) and therefore, more relevant per produced unit (Haas *et al.*, 2000; Halberg *et al.*, 2005). In the case considered here, where the focus is on GHG emissions related to organic wheat, the functional unit could be 1 kg of wheat at farmgate produced in Denmark. If the focus were wheat for bread, the protein content of the wheat should also be mentioned. If the wheat is used for pasta, other specific requirements might be mentioned. Therefore, the functional unit will differ depending on the aim of the study.

Goal. In this case, using organic wheat as a hypothetical example, setting the goal requires determining the differences in the GhG emissions between organic and conventional wheat grown in Denmark.

©Gustavo Fonseca de Almeida

Examining organic tomatoes

Scope. It requires defining the system boundaries, the procedure for co-product allocation, the methodology used which includes the choice of impact categories, and key assumptions and strategies for data collection and data quality.

System boundaries. Defining the scope of the study includes setting the system boundaries. For an agricultural product, this may be limited to the farmgate or it may include steps from processing to a regional distribution centre or even follow the product to the end user and disposal. With food, the end-user stage may differ with regard to how the food is handled in consumers' kitchens (e.g. boiled, roasted, fried or left raw). Several studies of agricultural products end at farmgate (e.g. Thomassen *et al.*, 2008; Halberg *et al.*, 2010; Casey and Holden, 2006) while others end at regional distribution centres (Hospido *et al.*, 2009; Williams *et al.*, 2008; Saunders *et al.*, 2006). Since the goal of the study of organic wheat is a comparative LCA of organic and conventional wheat production, it is not considered necessary to include life cycle steps after the farmgate, since the main differences take place at the farm level. however, if the aim of the study were to compare two different bread types, the functional unit might be 1 kg of a specific kind of bread and the milling and baking steps would thus be included in the analysis.

Allocation procedure. Defining the allocation procedure is also part of the scope definition. According to the ISO standards 14040 and 14044 that cover LCA, the allocation procedure – meaning the unit process cannot be divided into subprocesses – may consist of either (i) expanding the systems to include the additional functions related to the co-products or (ii) allocating according to, e.g. mass, economy or energy. In food production, and particularly livestock production, it is common to produce more than one product. In these cases, there is a need to allocate the environmental burden among different products. In principle, ISO standard 14044 (2006) recommends avoiding allocation, if possible, by dividing the unit process to be allocated into subprocesses, or expanding the product system to include the additional functions of the co-products. Otherwise, allocation for the system can be done in such a way that it reflects the physical properties or the relative economic values of co-products. There is a relationship between the choice of method – allocation or system expansion – and the choice of LCA approach – attributional or consequential (Nielsen *et al.*, 2003). The consequential approach, seeking to capture change in environmental impact as a consequence of actions, avoids co-product allocation by system expansion. The attributional approach deals with co-product allocation by partitioning the environmental impact related to the product using allocation factors based on mass, energy or economic value. Among all types of allocation listed, allocation based on economic value seems to be the preferred approach, since it reflects the underlying economic reasons for production (Jonasson and Sanden, 2004).

Small-scale organic farmers

Environmental impact categories. The environmental impact categories concerning the effect on climate change meaning global warming potential, and water and air pollution meaning eutrophication and acidification, are commonly used in agricultural LCAs, while effects on e.g. biodiversity or ecotoxicity due to pesticides are rarely assessed, due to methodological difficulties (e.g. Thomassen *et al.*, 2008; Halberg *et al.*, 2010). The implementation of socioeconomic impacts in LCA is still in its infancy (Griesshammer *et al.*, 2006; Jørgensen *et al.*, 2008). According to its aim, the only environmental impact category the organic wheat study considers is climate change. As mentioned earlier, there is a risk of sub-optimizing in relation to overall sustainability when only including one environmental impact category.

Data collection. The strategy for data collection and data quality will affect the representatives of the final results. This is discussed further in the text.

Inventory analysis

The life cycle inventory is typically the most time-consuming phase, since it includes data collection and treatment of data (Roy *et al.*, 2009). Site-specific data are needed from the production and processing stage, while other data on such as electricity and transport may be found in databases (Figure 14). Furthermore, emissions need to be estimated.

Understanding the N cycle is crucial, because the supply of N enables crop growth, but surplus N will always be emitted in some form, mainly N_2, NO, N_2O, NH_3 or NO_3^-, of which only N_2 has no negative environmental effects. Quantifying the N cycle requires knowledge of both the processes (e.g. N fixing rates, NH_3 emissions) and the activity data of farm practices.

Emissions, inputs and outputs are often estimated on a yearly basis. Such static annual time perspectives are commonly used in LCA. However, emissions related to one year's activities are not always restricted to the same year, which is the case for soil carbon emissions. Implementation of a more dynamic time perspective in LCA is still in its infancy (Levasseur *et al.*, 2010). The choice of whether the time perspective can be static or should be dynamic depends on the objectives and the studied objects. Another challenge is to obtain representative data based on several farms as a foundation for the analysis, as discussed further in the text.

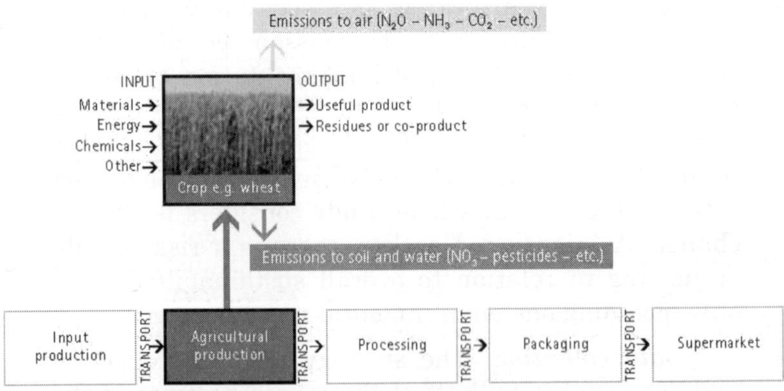

Figure 14

Illustration of the basic data requirements and emission estimates used for the LCA of agricultural products

The step "Agricultural production" is used as an example, but the same data requirements apply to each of the other steps ("Processing", "Packaging", etc.) in the product's life cycle if they are included in the analysis. Upstream processes (such as e.g. 'Input production') will always be included in the analysis.

In the case of organic wheat, data are needed on inputs (materials, energy, chemicals and other) and outputs (products, co-products) as illustrated in Figure 14. On the basis of this, the environmental production costs of the inputs can be estimated, using databases such as Ecoinvent (Ecoinvent Centre, 2011). Emissions, such as N_2O, NH_3, N leaching and soil C changes, also need to be estimated, using N balances and guidelines/modelling, such as the IPCC guidelines for N_2O (Intergovernmental Panel on Climate Change, 2006).

Impact assessment

The impact assessment implies characterization. In this step, emissions are assigned to the relevant impact categories, converted into the main unit used in the concerned impact category and aggregated with other relevant emissions within the same impact category. The primary emissions contributing to climate change are CO_2, CH_4 and N_2O. The impact category "global warming potential" is measured in CO_2 equivalents. Since CH_4 and N_2O also contribute to global warming, they need to be converted into CO_2 equivalents as well. The characterization factor describes the relative strengths of, for example Ch4 compared to CO_2 in a 100-year perspective.

Table 1 shows the characterization factors of some environmental impact categories that are commonly used in LCA. A characterization factor of 25 for CH_4 means that the CH_4 emissions should be multiplied by 25 to get the impact in CO_2 equivalents. Thus, 1 kg CH_4 has the same global warming impact in a 100-year perspective as 25 kg CO_2. The same principle goes for the conversion of all the emission elements into their relevant impact category. As stated in Table 1, the characterization factor of N_2O is 298, indicating that N2O is a powerful GHG. Thus, when the CO_2, CH_4 and N_2O emissions from the organic wheat production have been estimated, they need to be converted into CO_2 equivalents and factored into the total amount of CO_2 equivalents related to organic wheat production.

Table 1

Example of environmental impact categories used in LCA and the contributions from the main emissions

IMPACT CATEGORY	UNIT	CONTRIBUTING ELEMENTS	CHARACTERIZATION FACTORS
Land use	m^2	Land occupation	1 for all types of land use
Non-renewable energy	MJ	Non-renewable energy consumption	1
Global warming	CO_2 equivalents	CO_2	1
		CH_4	25
		N_2O	298
Acidification	SO_2 equivalents	SO_2	1
		NH_2	1.88
		NO_2	0.70
Eutrophication	NO_2 equivalents	NO_2	1
		PO_3	10.45
		NH_3	3.64
		NO_3	1.35

Source: Soloman *et al.* 2007; Wenezel *et al.,* 1997.

Interpretation

The fourth phase, interpretation, calls for an evaluation of the results, including sensitivity analysis and conclusions. The case study of organic soybeans in Box 1 (based on Knudsen *et al.,* 2010) illustrates the use of the LCA methodology described above.

Box 1

LCA of organic soybeans exported from China to Denmark

Goal and scope definition

Goal

The primary goal of the study is to identify the environmental hotspots in the product chain of organic soybeans produced in Jilin Province, China, and exported to Denmark. The secondary goal is to compare the environmental impacts in the production of organic soybeans with a conventional production in the Jilin Province.

Functional unit

The functional unit for the hotspot analysis: 1 tonne of organic soybean produced in China and delivered to Aarhus, Denmark. The functional unit for the comparison: 1 tonne of soybean produced in the case area in China leaving the farmgate.

System boundaries

The hotspot analysis (primary goal) studies the soybeans in all stages until delivery at the harbour in Denmark, including input production, farm stage (e.g. N_2O emissions), processing (e.g. drying, sorting, packing) and transport (including transport between all the steps and to final destination in Denmark). The comparison of organic and conventional soybeans, which is the secondary goal, only studies the soybeans until farmgate.

Impact categories

The impact categories considered are effects on climate change (global warming potential), eutrophication and acidification. Furthermore, non-renewable energy use and land use are reported in the study.

Data collection

The primary data were collected in the case study area, Jilin Province, China, where the soybeans exported to Denmark were produced. Farm data from 20 organic and 15 conventional farms were included in the study. The farm soils were mollisols and the main crops were soybeans (80 percent) and maize. A large number of the conventional farmers burn part of the soybean residue in the field, while the organic farmers either leave residue in the field or mix it with manure and forest soil and use it for compost.

Inventory analysis

The inputs and outputs from the soybean production were collected using questionnaires and conducting interviews with the farmers and the processing industry, to gather such information as the amounts of mineral fertilizer, compost, diesel, electricity and yields. The emissions related to the soybean production were estimated using mainly IPCC (2006) guidelines and nitrogen balances.

Impact assessment and interpretation

Hotspot analysis

The hotspot analysis of the organic soybeans produced in China and transported to Denmark found that the transport stage contributes 51 percent to the total GHG emissions from imported organic soybean followed by the farm stage (mainly N_2O emissions). The processing stage (including sorting and packaging) only contributes 11 percent (Figure 15).

Comparison of organic and conventional

The land-use requirements of the organic production of soybeans were 12 percent higher, whereas the other environmental impacts of organic soybeans at farmgate were approximately 50 percent less than the impacts for the conventional soybeans (Figure 16).

Sensitivity analysis

The sensitivity analysis showed that the estimated N_2O had a major impact on the results for the GHG emissions. The analysis also showed that including soil carbon changes in the life cycle assessment increased the difference in GHG emissions between organic and conventional soybeans.

Conclusion

The organic soybeans had a lower environmental impact with regard to non-renewable energy use, global warming, acidification and eutrophication potential per tonne produced compared to the conventional soybeans. The transport stage had a major impact (51 percent) on the GHG emissions from the imported organic soybeans to Denmark, followed by the farm stage (35 percent).

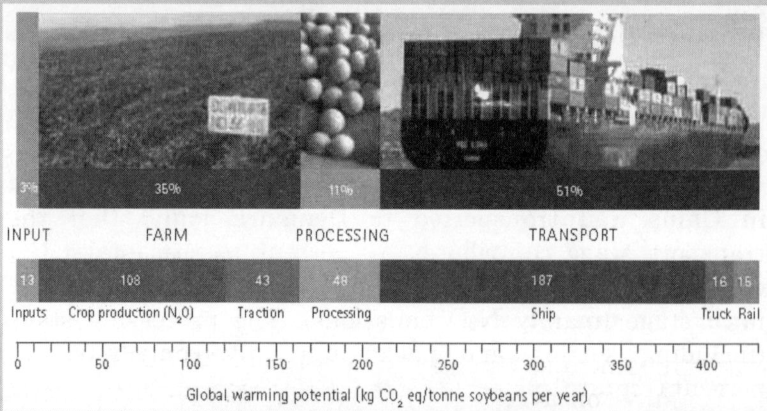

Figure 15
Greenhouse gas emissions from organic soybeans produced in Jilin province, China, and transported to the harbour of Aarhus, Denmark: a hotspot analysis

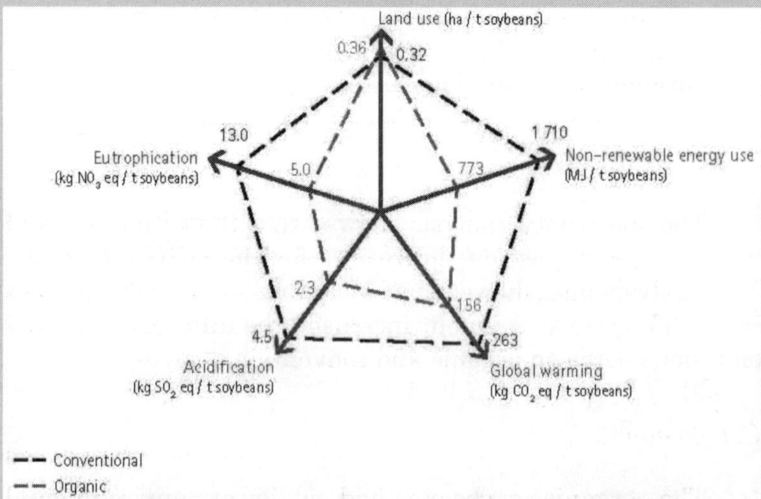

Figure 16
Environmental impacts from organic and conventional soybean produced in Jilin province, China

GREENHOUSE GAS EMISSIONS OF ORGANIC VERSUS CONVENTIONAL PRODUCTS

Assessing differences in greenhouse gas emissions between farming systems and agricultural products

A number of studies have evaluated the general environmental impacts of organic versus conventional products and farming systems, mainly in the European or North American context (e.g. Gomiero *et al.*, 2008; Mondelaers *et al.*, 2009). Other published studies have specifically evaluated the Australian (Wood *et al.*, 2006) and Canadian (Lynch, 2009) contexts. The overall conclusions have shown:

- organic matter – soils in organic farming systems have, on average, a higher content of organic matter (e.g. Mäder *et al.*, 2002; Fliessbach *et al.*, 2007; Mondelaers *et al.*, 2009);
- biodiversity – organic farming contributes positively to agro-biodiversity and natural biodiversity (e.g. Bengtsson *et al.*, 2005; Hole *et al.*, 2005; Mondelaers *et al.*, 2009);
- pesticides – organic agriculture minimizes the risk of conventional pesticide accidents and pollution, even though some substances such as copper are allowed in organic agriculture in some countries;
- GHG emissions – the conclusion is not that straight forward when assessing the impact of the organic farming system on GHG emissions and nitrate and phosphorous leaching; when expressed per production area, organic farming performs better than conventional farming for these impacts (e.g. Mondelaers *et al.*, 2009), but due to generally lower yields of organic farming, at least in developed countries, this positive effect expressed per unit product is less pronounced or not present at all (Mondelaers *et al.*, 2009).

The issue of GHG emissions is illustrated in Figure 17, which presents the results from a review of LCA studies that compared organic and conventional (Knudsen, 2011). The review found no general differences between the GHG embedded in organic and conventional product, with 20 studies above the line, where organic performs better than conventional and 8 studies below the line where organic performs worse. For plant products, only 2 of 16 studies showed organic performing worse than conventional agriculture.

Figure 17 also shows the relative importance of farming system (organic *vs.* conventional) and product type (plant *vs.* meat products) for GHG emissions. Beef had the highest values followed by lamb, pork, poultry and eggs. To reduce the GHG emissions related to the consumption of food, the replacement of meat by plant products means more than replacing conventional products with organic ones. Interestingly, it is often the case that eating organic is combined with eating less meat. A survey by a Danish supermarket found that the meat consumption by consumers who did not buy organic was twice as high (172 g meat per day) as the consumers who bought organic food (86 g meat per day) (FDB, 2010). Similar patterns might be found for canteens converting from conventional to organic food, due to higher prices of organic food and especially meat, implying a lowering of the meat proportion in the diets. These aspects are outside this work but must be considered in an overall interpretation of the impact of organic food production and consumption.

Soil carbon sequestration was generally not included in the LCAs presented in Figure 17 due to methodological limitations. however, since organic farming systems on average have a higher content of organic matter (Chapter 2 and Mondelaers *et al.*, 2009; Fliessbach *et al.*, 2007; Mäder *et al.*, 2002) this methodological gap needs to be solved and soil carbon sequestration included in LCAs. Interestingly, studies by Halberg *et al.* (2010) and Knudsen *et al.* (2010, 2011) indicated that the inclusion of estimated soil carbon changes in the sensitivity analysis widened the difference in GHG emissions per kg product between organic and conventional. however, there is a need to develop methodology for estimating and including soil carbon changes in LCA. As previously explained, traditionally it has not been included. Nevertheless, the results indicate that at least organic plant products might perform better than indicated in Figure 17, where most of the studies did not include soil carbon changes.

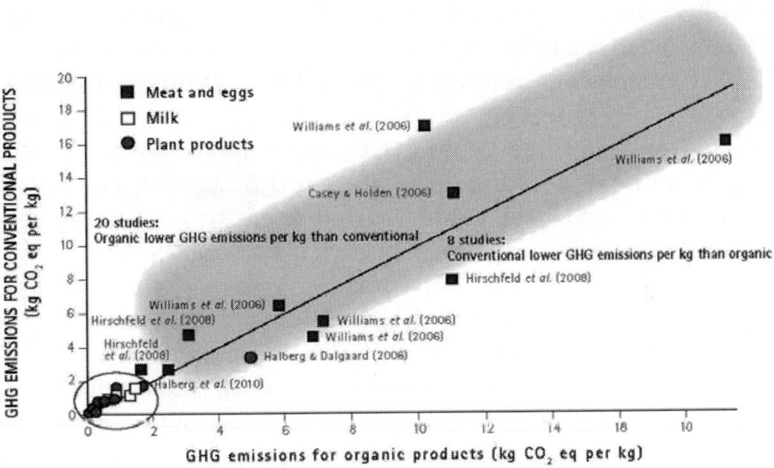

Figure 17.

Literature review of greenhouse gas emissions per kilogram of organic and conventional products

Organic perform better above the line and conventional perform better below the line. Idea after Niggli *et al.* (2008).

The upper graph contains the total number of LCA studies, whereas the lower graph is a zoom in on the studies of milk and plant products.

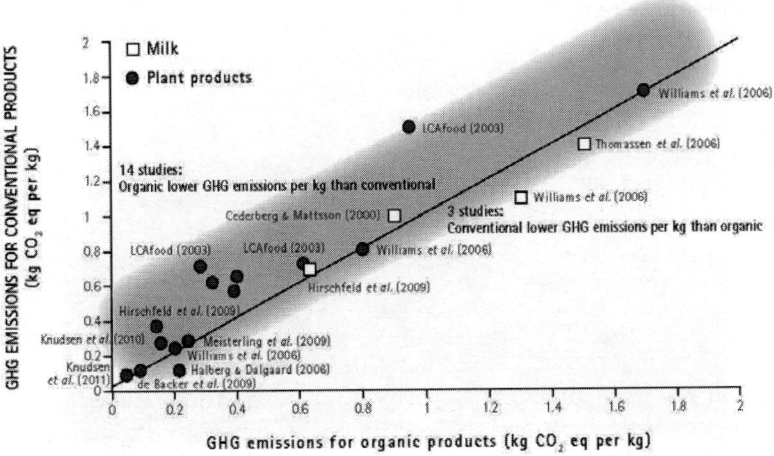

Differences in studies due to different farming practices or methodologies. It is apparent from Figure 17 (based on data in Annex 1), that different studies of the same product, such as milk, beef or wheat, do not always agree on the GHG emissions related to the product. The reported GHG emissions in Figure 17, related for example to organic wheat, varied from 0.14 to 0.80 kg CO_2 equivalent per kg wheat, organic milk values range from 0.63 to 1.50 kg CO_2 equivalent per kg milk, and beef values varied from 3.1 to 19.2 kg CO_2 equivalent per kg beef (see Annex 1). For organic wheat, the highest value of 0.8 kg CO_2 equivalent per kg wheat was for bread wheat, which needs high nitrogen inputs due to the baking quality – this partly explains the higher level of GHG emissions per kg grain. However, it does not explain the variation among the rest of the wheat studies. The variation in GHG estimates of 1 litre of organic milk was discussed by de Boer (2003) who concluded that the actual values of different LCA studies cannot be compared due to differences in methodology, and there is a need for further international standardization. Thus, organic and conventional products can only be compared within a case study (de Boer, 2003), which in fact is the case for the studies presented in Figure 17. The variation among LCA values for the same product is especially visible for organic beef, with the variation due to differences in production systems (dairy cattle vs. suckler beef), and to differences in methodology (e.g. with regard to allocation).

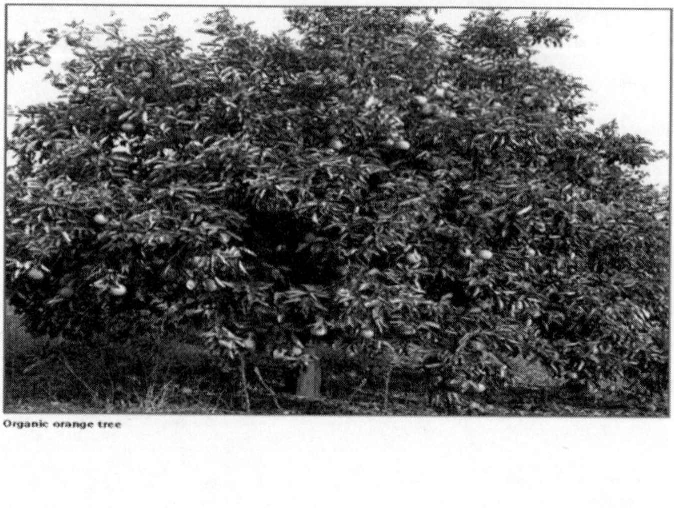

Organic orange tree

Differences within organic systems for the same product. As for comparisons between organic and conventional, differences within organic systems producing the same product can only be compared within the same case study, where the methodology is the same. however, few studies compare different organic farming systems producing the same product. Hirschfeld *et al.* (2008) conducted LCAs of both an average organic practice and an organic practice using climate-friendly cultivation methods to indicate how much climate impacts could be reduced. Knudsen *et al.* (2010, 2011) also found differences in GHG emissions among different organic orange production systems in Brazil. Differences in GHG emissions among different organic production systems of the same product can be used for benchmarking and point at mitigation options within organic farming systems.

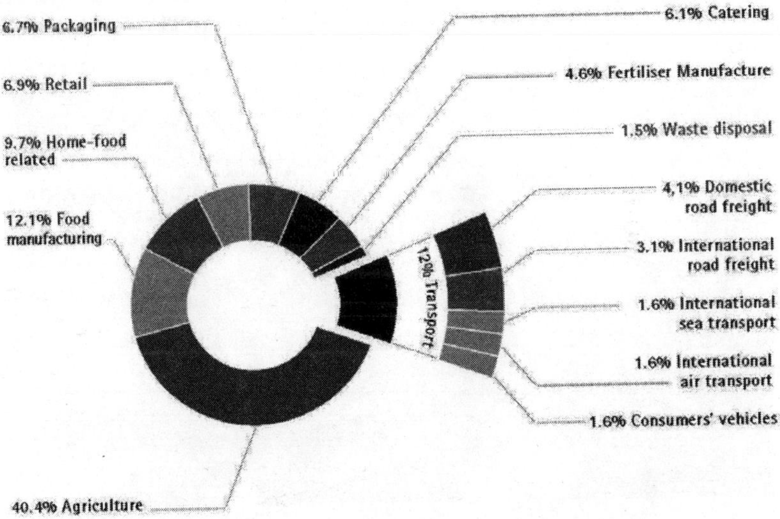

Figure 18

Greenhouse gas emissions from UK food consumption

Source: Oxfam, 2009

Major greenhouse gas contributions and mitigation options in organic food chains

Major contributions of GHG emissions for organic food and farming systems are mainly related to on-farm production. For traded plant products, transport is also a main contributor to GHG emissions related to the product, as shown by Knudsen (2011) and Knudsen *et al.* (2010), whereas the relative contribution is much smaller for meat.

Figure 18 shows an example of UK food consumption with the contribution to GhG emissions of every step in the food chain. It indicates that transport and processing, as well as on-farm production, are considerable hotspots, where relevant mitigation options would be beneficial. The main mitigation options at the farm level and for the food system are mentioned in Box 2.

Box 2

Climatic mitigation options in organic food chains

Farm level strategies

Energy:

- reduce field passes for operations,
- reduce pumped irrigation,
- minimize field crop drying,

- employ biodiesel and fuel efficient farm equipment,
- use crop rotations with a low energy requirement (perennial, N fixing, durable or grazed crops),
- increase plant and livestock production efficiency (C4 plants, farming systems design).

Nitrogen:

- increase efficient nitrogen utilization – optimizing crop rotations,
- Minimize N loss from fields, stables and manure storage.

Carbon:

- reduce methane emissions,
- avoid cultivation of peat soils,
- increase soil and vegetation carbon levels (use of manure, avoid crop residue burning, perennial grasslands, etc.).

Food system issues

- minimize transport of inputs and products, especially air and road transport,
- make more efficient and sustainable use of land,
- increase nutrient recycling from households to farms,
- minimize food waste,
- reduce meat consumption, increase pulse consumption instead,
- reduce consumption of highly processed high calorie foods,
- reduce packaging.

Source: MacRae, 2009

HOW TO PERFORM LIFE CYCLE ASSESSMENT IN COMPLEX AGRICULTURAL SYSTEMS

While LCA is recognized as the best tool for assessing the life cycle impacts of products (Finnveden *et al.*, 2009), especially with regard to GHG emissions, the LCA methodology also has some shortcomings. With regard to agricultural products, especially organic products derived from slightly more complex systems, several challenges are identified.

Interaction and interdependence between different components in the farming systems imply some challenges and allocation considerations, e.g. with regard to allocation aspects when including a green manure crop or catch crops in the crop rotation or estimating the environmental costs when using manure as a fertilizer.

Some aspects are still in the infancy of being developed and implemented in the global warming impact category, such as implementation of soil carbon changes and direct and indirect land use change. Not all impact categories are well covered in a typical LCA due to the need for further methodological development (Reap *et al.,* 2008). With regard to agricultural systems, impact categories regarding land use, including biodiversity and soil, are problematic and need to be improved (Finnveden *et al.,* 2009). However, the following only discusses those aspects and shortcomings related to global warming potential as a single impact category.

How to allocate and account for interactions in the farming systems

Absence of mineral fertilizer is one of the main differences between organic and conventional management practices. This makes organic farmers much more dependent than conventional farmers on crop rotation, especially because it provides nitrogen for the crops, and on livestock production systems, especially because they provide animal manure as a nitrogen source. In general, mixed-crop livestock systems are more frequent in organic farming due to the aim to connect crop and livestock production. Generally, the main nitrogen inputs to organic systems are derived from either green manure crops in the crop rotation or animal manure. The organic production systems that include green manure crops face the challenge of determining how to allocate the environmental burden from the green manure in the crop rotation to the other crops. Furthermore, arable crops' dependence on animal manure implies challenges on how to estimate the environmental impact of the production of animal manure, as illustrated in Figure 19.

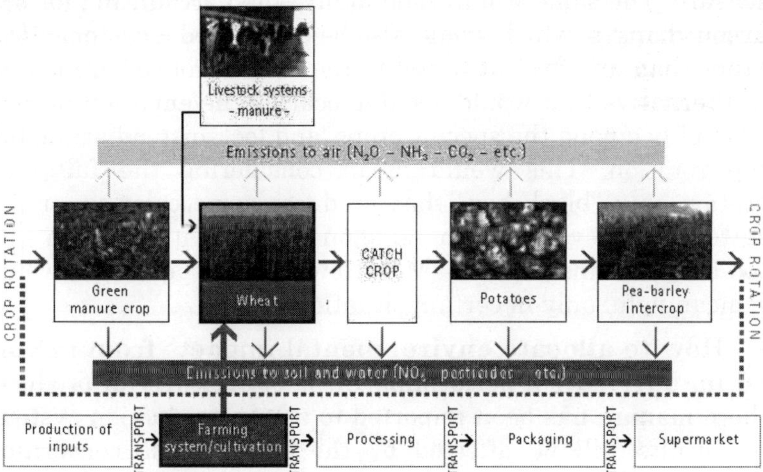

Figure 19

Illustration of the interactions in organic farming systems that needs to be accounted for in LCA of organic wheat

How to allocate environmental impacts or benefits between multiple products. In organic systems, crop rotation is central to the crops' nitrogen requirement and as a preventive measure for pest and disease management. With regard to nitrogen availability to the crops, green manure and catch crops might be included in the organic rotations. however, both green manure and catch crops have environmental impacts that cannot directly be related to a specific crop. Thus, the challenge is determining how the crops can share the burden of emissions or benefits of the green manure crops and catch crops. The environmental impacts (or benefits) of green manure, etc., might be allocated according to some biophysical relations such as the nitrogen utilization effect of the residuals, although it is hard to get reliable data to justify this allocation. The impact also could be allocated according to the economic value of the crops in the crop rotation. For now, the best path would seem to be recommending allocation of such impacts according to the area used to produce the different crops, because the effects could be considered "system" effects rather than the effect of one particular measure. The same would then apply when accounting for soil carbon changes, which would also be considered a system effect rather than an effect attached to the single crop. There is also an alternative that would avoid allocating the emission burden or benefits among the specific crops, and look instead at the full crop rotation. This would mean considering the full crop rotation as a "black box" that produces some calories, or dry matter or money, and then using one functional unit such as a food basket in mega joules (MJ). However, this approach would be meaningful only in certain situations.

How to allocate environmental impacts from imported manure. When performing LCAs of agricultural products where manure has been imported to a crop production system, the results will be affected by the assumptions concerning whether and how the environmental impacts from the manure is taken into account. This issue has been further described and discussed in Dalgaard and Halberg (2007) and Knudsen *et al.* (2010). In short, three approaches have been suggested for dealing with the manure issue:

- Dalgaard and Halberg (2007) suggested a consequential approach, in which the environmental costs of producing plant-available manure-N corresponds to the environmental costs of producing mineral fertilizer – the underlying assumption is that the manure could have replaced mineral fertilizer in another conventional field;
- Van Zeijts *et al.* (1999) suggested an LCA approach in which the emissions caused by storage, transport and application of animal manure are allocated according to the economic value of the manure;

- Audsley *et al.* (1997), and Jungbluth and Frischknecht (2007) discussed an approach that would include the environmental cost of producing manure-N corresponding to the environmental costs of producing the N in a green manure crop.

Overall, the approach could depend on the site-specific context of how the manure is used and regarded in the local society, and what kind of manure is used. Increasingly, animal manure is regarded as a precious source of fertilizer that replaces a certain amount of mineral fertilizer (for example corresponding to the amount of plant available N in the manure). In such cases, the first approach suggested by Dalgaard and Halberg (2007) of using the environmental costs of producing mineral fertilizer-N as a "shadow price" is relatively straight forward, although it is possible to discuss how many nutrients should be included (N alone or also P and K) and at which proportion (for example, assuming 60 percent of manure N content would replace fertilizer N). The animal manure used in the organic systems might either be derived from organic or conventional livestock production. When conventional animal manure is used in organic systems, the approach suggested by Dalgaard and Halberg (2007) of using the environmental production costs of mineral fertilizer as a shadow price is sound, if it takes into account the alternative value of the manure in the specific situation. Figure 20 illustrates the GHG emissions for organic wheat, depending on whether manure is regarded as a waste product or whether the environmental costs from the production and use of mineral fertilizer was used as a shadow price in different substitution rates. Contrary, when the organic systems are not importing manure from conventional livestock and thus, not related to the production of mineral fertilizer at all, this approach is less obvious. However, for now this methodology can be recommended since manure represents a valuable resource in the organic production system, and, when traded, needs a "currency".

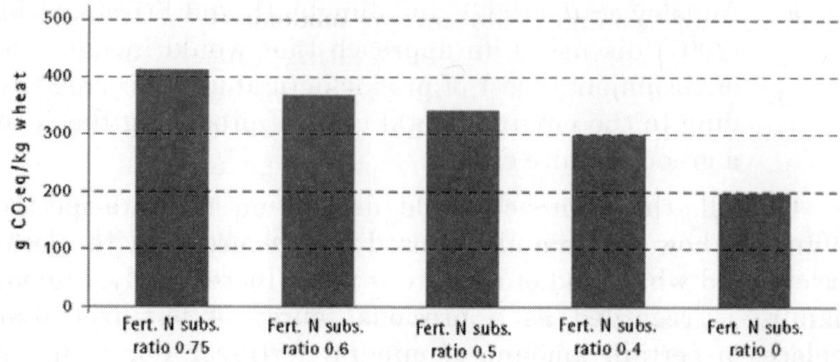

Figure 20

Greenhouse gas emissions (g CO_2 eq/kg wheat) of organic wheat relative to how the imported resource "manure" has been accounted for

How to account for carbon sequestration in life cycle assessment

The global warming potential category is well defined and central in LCAs. However, not all affected carbon is yet fully included in the calculations. The methods for estimating the changes in the organic carbon stocks in vegetation, soils and litter caused by agricultural production and including those in LCAs are still in the infancy of being discussed and implemented. Organic carbon changes have been discussed with regard to direct and indirect land use changes such as deforestation in relation to beef and soybeans from Brazil (Cederberg *et al.*, 2009) and bioenergy crops (e.g. Searchinger *et al.*, 2008). Soil carbon changes have gained attention in relation to comparisons of organic versus conventional production (Hörtenhuber *et al.*, 2010) and with regard to the use of straw for bioenergy (Levasseur *et al.*, 2010). Agreement on a simple and robust estimation method and a time horizon when including those changes in the LCA is crucial.

Soil carbon sequestration. As mentioned in Chapter 2, soils under organic management will often increase the level of soil organic carbon more than conventional management. The soil organic carbon changes can be estimated roughly using the IPCC Tier 1 simple methodology with default values (IPCC,

2006) or more complex soil carbon models. A more accurate methodological approach for estimating and including soil carbon changes in life cycle assessments has been suggested by Petersen *et al.* (submitted). They discuss the challenges of estimating the actual soil carbon change depending on the time perspective and development towards a new steady state level of soil organic carbon. The importance of choosing the time perspective use in the analysis is illustrated in Figure 21. Every farming practice implies a development towards a certain new steady state. however, the yearly changes towards this new steady state are larger in the beginning and then slow as the new steady state approaches. Thus, when choosing a time perspective of 20 years (which is used in IPCC, 2006), the changes and the effect of organic farming practices on soil carbon are more pronounced than if a time perspective of 100 years were used.

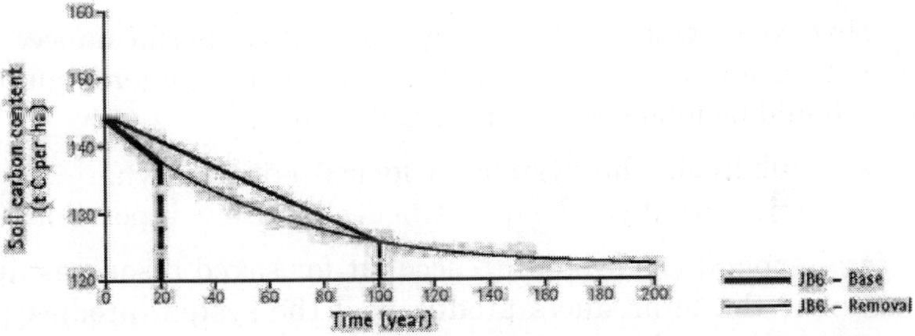

Figure 21
Illustration of the impact of the chosen time perspective when estimating soil carbon changes

Direct and indirect land-use change. There is an ongoing debate on direct and indirect land use change. This is especially relevant for bioenergy crops, soybeans and beef from Brazil and palm oil from Malaysia and Indonesia, which have a direct linkage to deforestation of rainforest and require new agricultural land for production. This is termed "direct" land-use change. "Indirect" land-use refers to the changes that are expected if the production systems in questions have changed demand for land compared to a previous situation which may consequently mean that land use moves to other places and indirectly causes change, e.g. deforestation. It could be argued that it is always relevant when using land for a purpose to include an indirect land-use change. However, many uncertainties and assumptions are involved in this argument and the methodology concerning land-use change is still in its infancy.

Recommendations and research needs

The LCA methodology is recommended for assessing the global warming impact related to production of organic products. However, the transfer of resources within the organic system makes it a very complex system to handle in the assessment of global warming impact. While ISO guidelines provide overall guidelines and should be followed for scientific purposes and internal decision making, there is still a need for further development of LCA methodologies for complex systems such as organic agriculture.

However, when evaluating the LCA performance of products from complex systems, the following general guidelines should be followed in prioritized order: 3

- subdivide the system in logical entities, where input and output can be quantified separately, if possible,
- expand the system to account for saved resources due to the bi-products produced, if the system produces a main product and a biproduct,
- make a biophysical modelling, if more than one main product is produced and it is meaningful to model the biophysical relations for each main product, its use of resources and responsibility for emissions,
- allocate according to mass or economic value.

Since certain schemes used for reporting of GHG emissions related to different product were already developed, it may be necessary to adhere to these for communication purposes. Some aspects, very important in organic production, should however be given further attention and need methodological development in order to establish well accepted procedures.

Interactions in cropping systems. The interactions and synergies in the organic cropping systems – such as the inclusion of green manure – and the fact that crops are included in a purposely chosen cropping pattern that aims to optimize the entire system, rather than the individual crop represent a particular challenge. A preliminary recommendation is that all emissions are allocated either equally on other crops in the crop rotation or based on the economic value of the outputs from the system, since the cropping pattern is probably constructed in a way that reflects the economic performance of the system. This recommendation however, needs more reflections based on studies of experimental and practical situations.

Manure and organic fertilizer. In the case of use of manure imported from outside the system under consideration, it is recommended to expand the system to take into account the indirect impact of using this resource, unless it is well documented that the manure or other organic fertilizer input are, in fact, waste products that would not be used otherwise. If the manure could have substituted as fertilizer elsewhere, this should be accounted for in the assessment following the principles described in the previous sections. Alternatively, if the manure could have been used in another organic production system, this also should ideally be taken into account. This may however, be very complex in practise, and as a proxy it is recommended to use an "exchange rate" for manure, based on the substitution rate for fertilizer in that area, or the alternative ("shadow value") costs of producing an equivalent effect via green manure production. There is need here to define a common agreed procedure.

Soil carbon sequestration. It is recommended to take soil carbon sequestration into account in LCA of organic products, following a transparent method that accounts for the decay and emissions of soil carbon and the time perspective. A 20-year perspective is most commonly used to estimate the GHG effect of soil carbon changes, but a 100-year perspective should be shown at least for sensitivity analysis. more research is needed to account for soil carbon dynamics as a consequence of cropping systems. The present empirical work on estimating soil carbon changes should be taken into account, but it is also recommended to initiate research to consider if and how to take the above-ground carbon sequestration into account in organic production systems. Presently, this carbon sequestration is taken into account when it results in a product exported from that mixed system. Since organic systems in many situations are more diverse than conventional systems, it is important to acknowledge and to take such effects into account. For example, if the systems include production of woods which are not harvested annually, this production needs to be included when analyzing the outcome of the assessment. It also may be relevant to develop a method similar to the soil model that accounts for the temporary sequestration, even if the C in the wood ultimately is released to the atmosphere again after a span of years. Furthermore, carbon sequestration and emissions from direct and indirect land use change to be included in LCAs are major issues, being hotly debated and investigated at the moment. Thus, at this point, land use change is not well implemented in LCA's due to methodological challenges, and more research is needed.

STATUS ON INITIATIVES AND DATA REQUIREMENTS OF LIFE CYCLE ASSESSMENT

Existing initiatives to assess life cycle assessment of products

Organic orange flower and fruit

In order to increase the transparency of environmental concerns related to products and harmonize the LCA methodology used for this, a number of initiatives have been taken. The importance of harmonization of the LCA methodology, as introduced in this Chapter, recognizes that differences in results for the same product might be due to differences in methodologies, rather than differences in farming systems. The following presents the most important attempts to harmonize LCA methodologies. ISO standards for life cycle assessment were developed and described in ISO 14040 and 14044. Since 2008, ISO has developed standards for quantifying (ISO 14067-1) and communicating (ISO 14067-2) the carbon footprint of products. These new standards are largely based on the existing ISO standards for LCAs (ISO 14040/44) and environmental labels and declarations (ISO 14025). The final standard is expected to be published in early 2012.

- PAS 2050, published in 2008, specifies the assessment of the life cycle for GHG emissions of goods and services (PAS 2050, 2008). This guideline, focusing specifically on GHG emissions, was initiated by Carbon Trust (a public nonprofit company) and UK Department for Environment, Food and Rural Affairs in 2007 and is hosted by the British Standards Institution. PAS 2050 builds on the existing ISO standard (ISO 14040/44) by specifying requirements for the assessment of the life cycle GHG emissions of products. The Carbon Reduction Label is a carbon footprint labelling system run by Carbon Trust that uses the PAS 2050 standard.
- EU Joint Research Centre for Environment and Sustainability is developing an International Reference Life Cycle Data System (ILCD), because the ISO 14040/44 standards leave the practitioner with a range of choices that can affect the results. The ILCD consists primarily of the ILCD handbook, published in 2010, and the ILCD data network which is still being developed. The ILCD handbook provides detailed guidance on all the steps required to conduct an LCA.

Its main goal is to ensure quality and consistency of life cycle data, methods and assessments.

Manure spreader

Data requirements and greenhouse gas emission estimates for LCA-based certification

The basic data requirement for assessing GHG of organic products, as described in Chapter 2, may depend on the methodology used. Basically the IPCC guidelines can be used in considering the need for establishing emission factors in different situations. however, the question is whether the guidelines offer the precision needed by developing countries, since much of the work in establishing emission factors took place under different conditions. It is important to know how emissions are estimated, since N_2O emissions for plant products and CH_4 emissions for animal products have major impact on the results.

Data sources for nitrogen losses include IPCC emission factors (EFs) for N_2O or national inventory EFs for ammonia. The IPCC Tier 1 approaches are simple methods using default values, Tier 2 approaches are similar, but with country specific emission factors and other data, while Tier 3 are more complex approaches, possibly models, but compatible with lower Tiers. Tier 2 or 3 might be preferred to the IPCC Tier 1 approaches, because Tier 1 has coarse, first-order approximations. So, they do not reliably reflect local manure management practices (whether non-organic or organic). The lack of refinement for Tier 1 EFs could easily mask actual differences in emissions between systems. On the other hand, the variation caused by different modelling approaches in different studies brings in an extra cause of variation when comparing different studies. So far, the Tier 1 approach of the IPCC 2006 guidelines is a generally accepted method for estimating N_2O and CH_4 emissions for the LCA. With regard to livestock and manure, the IPPC (2006) EFs for N_2O and CH_4 together with NH_3 EFs from national inventories form a basis for emissions, but practices such as composting are generally not well characterized.

Another aspect is the requirement for input-output data. If the data collection and assessment take place at a specific farm or group of farms, the question arises of how representative these farms are for the entire sector in that area. Conventional agriculture in Europe can rely on appropriate databases such as the Farm Accountancy Data Network (FADN). However, organic systems are not well represented in those statistics due to the limited size of the organic sector. Thus, LCA studies on organic products face a major challenge due to the limited availability of data from a large number of farms representing organic production, especially in developing countries where organic production is just emerging.

Existing databases of organic products evaluated with LCA methodology

Only a limited number of LCA studies on organic products have been published in peer-reviewed journals. most of these studies have included a comparison of organic and conventional agriculture, as shown in Figure 17. most of the values presented in Figure 17 are from reports (Halberg *et al.*, 2006; Hirshfeld *et al.*, 2008; Williams *et al.*, 2006) or databases (LCAfood, 2003). This also illustrates that the small size of the organic sector presents challenges in securing representative data and providing enough incentive for developing a consistent LCA methodology with regard to the specific requirements of the organic products.

Two main databases that provide data on organic products represent organic production in Denmark (LCAfood, 2003) and Switzerland (Ecoinvent Centre, 2011). In addition to this, there is a French LCA initiative on organic food and agricultural products. The most established LCA database is Ecoinvent, owned and managed by a consortium of Swiss research organizations. The database is under continuous development, but only made available to users commercially (Ecoinvent Centre, 2011).

The ILCD data network currently has limited data availability but it eventually will provide an open-access database. The European Commission is requiring all new research projects funded by the EU that are working with LCA case studies to use the ILCD handbook and deliver ILCD compatible datasets.

©Marie Trydeman Knudsen

Carbon rich soil in organic coffee plantation

SUMMARY AND NEXT STEPS

Participants of the RTOACC workshop were able to draw from their discussions and from the input of guest speakers and synthesize a set of conclusions that can be used to guide future activities concerning LCAs and other activities that seek to identify and quantify the potential contributions of organic agriculture to climate change mitigation. These include:

- LCA is the best tool for measuring GHG emissions related to agricultural products,
- there is a risk of over simplification when focusing on climate change as a single environmental impact category,
- farm production and transport (at least for plant products) are important hotspots for agricultural products,
- studies have shown no remarkable difference in GHG emissions between organic and conventional but, traditionally, soil carbon changes have not been included – which can have a major impact, especially for plant products,
- the challenges of LCA of organic products need to be addressed, such as accounting for carbon sequestration and interactions in farming systems, as well as including the environmental costs of manure,
- attempts should be made to secure a consistent LCA methodology for agricultural products, including organic products.

ANNEX

LITERATURE REVIEW OF GREENHOUSE GAS EMISSIONS PER KILOGRAM ORGANIC VERSUS CONVENTIONAL AGRICULTURAL PRODUCT AT FARMGATE

PRODUCT	GHG EMISSIONS PER KG PRODUCT AT FARMGATE (KG CO_2.eq/kg)			REFERECES
	Conventional	Organic	Ratio of Organic to conventional	
MEAT				
Beef, UK	16.0	19.2	1.2	Williams *et al.* (2006)
Suckeler beef, live weight	13.0	11.1	0.9	Casey & Holden (2006)
Beef, dairy cattle, Germany	4.8	3.1	0.6	Hirschfeld *et al.* (2008)
Fattening bull, dairy, Germany	7.9	11.0	14	Hirschfeld *et al.* (2008)
Pig meat, DK	2.7	2.5	0.9	Halberg *et al.* (2010)
Pig meat, UK	6.4	5.8	0.9	Williams *et al.* (2006)
Pig meat, Germany	2.72	1.7	0.6	Hirschfeld *et al.* (2008)
Poultry, UK	4.6	6.9	1.5	Williams *et al.* (2006)
Sheep, UP	17.0	10.2	0.6	Williams *et al.* (2006)
Eggs, UK	5.5	7.2	1.3	Williams *et al.* (2006)
DAIRY				
Milk, Germany	0.70	0.63	0.9	Hirschfeld *et al.* (2008)

PRODUCT	GHG EMISSIONS PER KG PRODUCT AT FARMGATE (KG CO_2.eq/kg)			REFERECES
	Conven-tional	Organic	Ratio of Orga-nic to conven tional	
Milk, The Netherlands	1.4[1]	1.5[1]	1.1	Thomassen *et al.* (2008)
Milk, UK	1.1	1.3	1.2	Williams *et al.* (2006)
Milk, Sweden	1.0[2]	0.9[2]	0.9	Cederberg & Mattsson (2000)
FRUIT/ VEGETABLES				
Oranges, Brazil	0.11	0.08	0.8	Knudsen *et al.* (2011)
Leeks, Belgium	0.094	0.044	0.5	de Backer *et al.* (2009)
Potatoes, UK	0.24	0.2	0.9	Williams *et al.* (2006)
Carrot, UK	0.12	0.21	1.7	Halberg *et al.* (2006)
Tomatoes, greenhouse, DK	3.45	4.96	1.4	Halberg *et al.* (2006)
AGRICULTURAL CROPS				
Soybeans, China				Knudsen *et al.* (2011)
Wheat, USA				Meisterling *et al.* (2009)
Wheat, Germany				Hirschfeld *et al.* (2008)
Bread wheat, UK				Williams *et al.* (2006)
Wheat, DK				LCAfood (2003)
Oilseed rape, UK				Williams *et al.* (2006)
Oilseed rape, DK				LCAfood (2003)
Winter barley, DK				LCAfood (2003)
Spring barley, DK				LCAfood (2003)
Oat, DK				LCAfood (2003)
Rye, DK				LCAfood (2003)

[1] per kg fat protein corrected milk (FPCM)

[2] per kg energy corrected milk (ECM)

REFERENCES

Audsley, A., Alber, S., Clift, R., Cowell, S., Crettaz, P., Gaillard, G., Hausheer, J., Jolliet, O., Kleijn, R., Mortensen, B., Pearce, D., Roger, E., teulon, H., Weidema, B. & van Zeijts H. 1997. *Harmonization of environmental life cycle assessment for agriculture.* Final Report for Concerted Action AIR3-CT94-2028. Silsoe Research Institute, Silsoe, United Kingdom. 101 p.

Bengtsson, J., Ahnström, J. & Weibull A.C. 2005. The effects of organic agriculture on biodiversity and abundance: A meta-analysis. *Journal of Applied Ecology*, 42: 261–269.

Casey, J.W. & Holden, N.M. 2006. Greenhouse gas emissions from conventional, agrienvironmental scheme and organic Irish Suckler-Beef. *Journal of Environmental Quality*, 35: 231–239.

Cederberg, C. & Mattsson, B. 2000. Life cycle assessment of milk production — A comparison of conventional and organic farming. *Journal of Cleaner Production*, 8: 49–60.

Cederberg, C., Meyer, D., & Flysjö, A. 2009. Life cycle inventory of greenhouse gas emissions and use of land and energy in Brazilian beef production. SIK report No. 792. 67 p. (available at http://www.sik.se/archive/pdf-filer-katalog/SR792.pdf)

Dalgaard, R. & Halberg, N. 2007. how to account for emissions from manure? Who bears the burden? In: **Proceedings from the 5th International Conference 'LCA in Foods'**, 25–26 April 2007, Gothenburg, Sweden.

de Backer, E., Aertsens, J., Vergucht, S. & Steurbaut, W. 2009. Assessing the ecological soundness of organic and conventional agriculture by means of life cycle assessment (LCA). A case study of leek production. *British Food Journal*, 111 (10): 1028–1061.

de Boer, I.J.M. 2003. Review. Environmental impact assessment of conventional and organic milk production. *Livestock Production Science*, 80: 69–77.

de Koeijer, T.J., Wossink, G.A.A., Struik, P.C. & Renkema, J.A. 2002. measuring agricultural sustainability in terms of efficiency: the case of Dutch sugar beet growers. *Journal of Environmental Management*, 66: 9–17.

Ecoinvent Centre. 2011. **Ecoinvent Database v. 2.2.** Swiss Centre for Life Cycle Inventories St-Gallen, Switzerland (available at http://www.ecoinvent.org/database/).

FDB. 2010. Økologiske forbrugere belaster klimaet mindre [Danish]. Organic consumers have less climate impact (available at http://fdb.dk/nyhed/%C3%B8kologiske-forbrugerebelaster-klimaet-mindre).

Finnveden, G., Hauschild, M.Z., Ekvall, T., Guninée, J., Heijungs, R., Hellweg, S., Koehler, A., Pennington, D. & Suh, S. 2009. Review. Recent developments in life cycle assessment. *Journal of Environmental Management*, 91:1–21.

Fliessbach, A., Imhof, D., Brunner, T. & Wüthrich, C. 1999. Tiefenverteilung und zeitliche Dynamik der mikrobiellen Biomasse in biologisch und konventionell bewirtschafteten Böden. Regio Basiliensis 3, 253–263.

Fliessbach, A., Oberholzer, H.R., Gunst, L. & Mäder, P. 2007. Soil organic matter and biological soil quality indicators after 21 years of organic and conventional farming. *Agriculture, Ecosystems and Environment*, 118: 273–284.

Foster, C., Green, K., Bleda, M., Dewick, P., Evans, B., Flynn, A. & Mylan, J. 2006. Environmental impacts of food production and consumption. A report to the Department for Environment, Food and Rural Affairs. Manchester Business School. DEFRA, London, UK 199 p.

Government of Alberta (GoA). 2010. Quantification protocol for agricultural nitrous oxide emissions reductions, specified gas emitters regulation. Government of Alberta, Canada, (available at: http://carbonoffsetsolutions.climatechange-central.com/offsetprotocols/approved-alberta-protocols).

Gomiero, T., Paoletti, M.G. & Pimentel, D. 2008. Energy and environmental issues in organic and conventional agriculture. *Critical Reviews in Plant Sciences*, 27: 239–254.

Griesshammer, R., Benoît, C., Dreyer, L.C., Flysjö, A., Manhart, A., Mazijn, B., Méthot, A.L. & Weidema, B.P. 2006. Feasibility study: integration of social aspects into LCA. UNEPSETAC Life Cycle Initiative, Paris, France.

Haas, G., Wetterich, F. & Geier, U. 2000. Life cycle assessment framework in agriculture on the farm level. *International Journal of Life Cycle Assessment*, 5(6): 345–348.

Halberg, N., Dalgaard, R. & Rasmussen M.D. 2006. miljøvurdering af konventionel og økologisk avl af grøntsager – Livscyklusvurdering af produktion i væksthuse og på friland: Tomater, agurker, løg, gulerødder. (Life Cycle Assessment of Conventional and organic vegetable production). Arbejdsrapport fra miljøstyrelsen no. 5. (available at http://orgprints.org/13085/)

Halberg, N., Hermansen, J.E., Kristensen, I.S., Eriksen, J., Tvedegaard, N. & Petersen, B.M. 2010. Impact of organic pig production systems on CO_2 emission, C sequestration and nitrate pollution. *Agronomy for Sustainable Development*, 30: 721–731.

Halberg, N., van der Werf, H.M.G., Basset-Mens, C., Dalgaard, R. & de Boer, I.J.M. 2005. Environmental assessment tools for the evaluation and improvement of European livestock production systems. *Livestock Production Science*, 96: 33–50.

Hamilton, K., Peters-Stanley, M., & Marcello, T. 2010. **Building bridges: state of the voluntary carbon markets 2010**. Ecosystem marketplace, Washington, USA & Bloomberg New Energy Finance, New York, USA (available at http://www.forest-trends. org/documents/files/doc_2433.pdf).

Häni, M. 2010. **Current data availability of soil carbon stocks in Switzerland neighboring regions: a meta-analysis**. master Thesis. Swiss Federal Institute of Technology, Zurich (ETH) and Research Institute of Organic Agriculture (FiBL), Zürich, Switzerland.

Hirschfeld, J., Weiß, J., Preidl, M. & Korbun, T. 2008. Klimawirkungen der Landwirtschaft in Deutschland (The impact of German agriculture on the climate). Discussion paper of the Institut für ökologische Wirtschaftsforschung (IÖW), 189/08. 42 p.

Hole, D.G., Perkins, A.J., Wilson, J.D., Alexander, I.H., Grice, P.V. & Evans, A.D. 2005. Does organic farming benefit biodiversity? *Biological Conservation*, 122: 113–130.

Hörtenhuber, S., Lindenthal, T., Amon, B., Markut, T., Kirner, L. & Zollitsch, W. 2010. Greenhouse gas emissions from selected Austrian dairy production systems – model calculations considering the effects of land use change. *Renewable Agriculture and Food Systems*, 25: 316–329.

Hospido, A., Milà I Canals, L., McLaren, S., Truninger, M., Edwards-Jones, G. & Clift, R. 2009. The role of seasonality in lettuce consumption: A case study of environmental and social aspects. *International Journal of Life Cycle Assessment*, 14 (5): 381–391.

International Fertilizer Industry Association (IFA). 2009. the Global "4R" nutrient stewardship framework. Developing fertilizer best management practices for delivering economic, social and environmental Benefits. IFA, Paris, France 10 pp.

International Federation of Organic Agriculture Movements (IFOAM). 2005. **Principles of organic agriculture**. IFOAm. 4 p. (available at http://www.ifoam.org/organic_facts/principles/pdfs/IFOAm_FS_Principles_forWebsite.pdf).

Intergovernmental Panel on Climate Change (IPCC). 2006. **Guidelines for national greenhouse gas inventories**. IPCC National Greenhouse Gas Inventories Programme (available at http://www.ipcc-nggip.iges.or.jp/public/2006gl/index.html).

Solomon, S., Qin, D., Manning, M., Chen, Z., Marquis, M., Averyt, K.B., Tignor M. & Miller, H.L. (eds.) 2007. Climate change 2007: The physical science basis. contribution of working group I to the fourth assessment report of the Intergovernmental Panel on Climate Change. Cambridge University Press, Cambridge, United Kingdom and New York, NY, USA, 996 p.

International Organization for Standardization (ISO). 2006. Environmental management - Life cycle assessment - Requirements and guidelines. ISO 14044, Geneva, Switzerland.

Jonasson, K. & Sanden, B. 2004. time and scale aspects in life cycle assessment of emerging technologies-case study on alternative transport fuels. Centre for Environmental Assessment of Product and material Systems (CPm)-report 2004: 6, Chalmers University of Technology, Göteborg, Sweden (available at http://lifecyclecenter.se/wordpress/wpcontent/uploads/2011/01/2004_6.pdf).

Jørgensen, A., Le Bocq, A. & Hauschild, M.Z. 2008. methodologies for social life cycle assessment – A review. *International Journal of Life Cycle Assessment*, 13: 96–103.

Jungbluth, N. & Frischknecht, R. 2007. Life cycle assessment of imported agricultural products – Impacts due to deforestation and burning of residues. In: **Proceedings from the 5th International Conference 'LCA in Foods'**, 25–26 April 2007, Gothenburg, Sweden.

Knudsen, M.T. 2011. Environmental assessment of imported organic products – focusing on orange juice from Brazil and soybeans from China. PhD thesis. Faculty of Life Sciences, University of Copenhagen & Faculty of Agricultural Sciences, Aarhus University, Denmark. ISBN 978-87-7611-408-4. 146 p.

Knudsen, M.T., de Almeida, G.F., Langer, V., de Abreu, L.S. & Halberg, N. 2011. Environmental assessment of organic juice imported to Denmark: A case study on oranges (*Citrus sinensis*) from Brazil. *Organic Agriculture*, 1:167–185.

Knudsen, M.T., Yu-Hui, Q., Yan, L. & Halberg, N. 2010. Environmental assessment of organic soybean (*Glycine max.*) imported from China to Denmark: A case study. *The Journal of Cleaner Production*, 18: 1431–1439.

Lal, R. 2004. Soil carbon sequestration impacts on global climate change and food security. *Science*, 304: 1623–1627.

LCAfood. 2003. **LCA food database** (available at http://www.lcafood.dk).

Leifeld, J. & Fuhrer, J. 2010. Organic farming and soil carbon sequestration: what do we really know about the benefits? *Ambio*, 39(8): 585–99.

Levasseur, A., Lesage, P., Margni, M., Deschenes, L. & Samson, R. 2010. Considering time in LCA: Dynamic LCA and its application to global warming impact assessments. *Environmental Science and Technology*, 44: 69–74.

Lynch, D. 2009. Environmental impacts of organic agriculture: A Canadian perspective. *Canadian Journal of Plant Science*, 89(4): 621–628.

MacRae, R. 2009. **Comparing energy use and GHG mitigation potentials in organic vs. conventional farming systems**. A discussion paper for the Organic Agriculture Centre of Canada, Truro, Nova Scotia, Canada. 33 p. (available at http://www.organicagcentre.ca/ Docs/Discussion Papers09/Energy_Use_rmc.pdf).

Mäder, P., Fliessbach, A., Dubois, D., Gunst, L., Fried, P. & Niggli, U. 2002. Soil fertility and biodiversity in organic farming. *Science*, 296: 1694–1697.

Meisterling, K., Samaras, C. & Schweizer, V. 2009. Decisions to reduce greenhouse gases from agriculture and product transport: LCA case study of organic and conventional wheat. *Journal of Cleaner Production*, 17: 222–230.

Mondelaers, K., Aertsens, J. & Van Huylenbroeck, G. 2009. A meta-analysis of the differences in environmental impacts between organic and conventional farming. *British Food Journal*, 111: 1098–1119.

Nielsen, P., Nielsen, A., Weidema, B., Dalgaard, R. & Halberg, N. 2003. LCA Food Data Base. In: Denmark. http://www.lcafood.dk.

Niggli, U., Schmid, H. & Fliessbach, A. 2008. Organic Farming and Climate Change. International Trade Centre (ITC), Geneva, Switzerland 30 p.

Oxfam. 2009. 4-a-week: Changing food consumption in the UK to benefit people and planet. *Oxfam GB Briefing Paper*, march 2009.

Publicly Available Specification (PAS) 2050. 2008. Specification for the assessment of the life cycle greenhouse gas emissions of goods and services. British Standards Institution (BSI). 37 p. (available at http://www.bsigroup.com/upload/ Standards%20 &%20Publications/Energy/PAS2050.pdf)

Petersen, B.M., Knudsen, M.T., Hermansen, J.E. & Halberg, N. (submitted). A methodological approach to include soil carbon changes in life cycle assessments.

Reap, J., Roman, F., Duncan, S. & Bras, B. 2008. A survey of un-resolved problems in life cycle assessment. Part 2: Impact assessment and interpretation. *International Journal of Life Cycle Assessment*, 13: 374–388.

Rethemeyer, J., Kramer, C., Gleixner, G., John, B., Yamashita, T., Flessa, H., Andersen, N., Nadeau, M.J. & Grootes, P.M. 2005. Transformation of organic matter in agricultural soils: Radiocarbon concentration versus soil depth. *Geoderma*, 128: 94–105.

Roy, P., Nei, D., Orikasa, T., Xu, O., Okadome, H., Nakamura, N. & Shiina, T. 2009. A review of life cycle assessment (LCA) on some food products. *Journal of Food Engineering*, 90: 1–10.

Sanderman, J. & Baldock, J.A. 2010. Accounting for soil carbon sequestration in national inventories: a soil scientist's perspective. *Environmental Research Letters*, 5: 1–6.

Saunders, C., Barber, A. & Taylo, R.G. 2006. Food miles – comparative energy/emissions performance of New Zealand's Agriculture Industry. *Research Report No. 285*. Agribusiness & Economics Research Unit, Lincoln University, New Zealand. 119 p.

Scialabba, N.E.H. & Müller-Lindenlauf, M. 2010. Organic agriculture and climate change. *Renewable Agriculture and Food Systems*, 25(2): 158–169.

Searchinger T., Heimlich, R., Houghton, R.A., Dong, F., Elobeid, A., Fabiosa, J, Tokgoz, S, Hayes, D. & Yu, T-H. 2008. Use of U.S. croplands for biofuels increases greenhouse gases through emissions from land-use change. *Science*, 319: 1238-1240.

Smith, P., Martino, D., Cai, Z., Gwary, D., Janzen, H., Kumar, P., McCarl, B., Ogle, S., O'Mara, F., Rice, C., Scholes, B., & Sirotenko, O. 2007. Agriculture. In: B. metz, O.R. Davidson, P.R. Bosch, R. Dave, & L.A. Meyer (eds). Climate Change 2007: Mitigation. Contribution of Working Group III to the Fourth Assessment Report of the Intergovernmental Panel on Climate Change, Cambridge University

Press, Cambridge, United Kingdom and New York, NY, USA (available at http://www.ipcc.ch/publications_ and_data/ar4/ wg3/en/contents.html).

Thomassen, M.A., van Calker, K.J., Smits, M.C.J., Iepema, G.L. & de Boer, I.J.M. 2008. Life cycle assessment of conventional and organic milk production in the Netherlands. *Agricultural Systems*, 96: 95–107.

van der Werf, H.M.G., Tzilivakis, J., Lewis, K. & Basset-Mens C. 2007. Environmental impacts of farm scenarios according to five assessment methods. *Agriculture Ecosystems and Environment*, 118: 327–338.

van Zeijts, H., Leneman, H. & Sleeswijk A.W. 1999. Fitting fertilisation in LCA: Allocation to crops in a cropping plan. *Journal of Cleaner Production*, 7:69-74.

Voluntary Carbon Standard (VCS) 2010. Quantifying N_2O emissions reductions in US agricultural crops through N-Fertilizer rate reduction. Voluntary Carbon Standard proposed methodology (available athttp://www.v-c-s.org/methodologies/quantifyingn2o-emissions-reductions-us-agricultural -crops-through-n-fertilizer-rate-0).

Wenzel, H., Hauschild, M. & Alting, L. 1997. Environmental assessments of products – Vol. 1: Methodology, tools and case studies in product development. Chapman and hall, London, UK.

Williams, A.G., Audsley, E. & Sandars, D.L. 2006. Determining the environmental burdens and resource use in the production of agricultural and horticultural commodities. In: **Main Report, Defra Research Project IS0205**. Cranfield University and Defra, UK (available at http://www.defra.go.uk).

Williams, A.G., Pell, E., Webb, J., Tribe, E., Evans, D., Moorhouse, E. & Watkiss P. 2008. Comparative life cycle assessment of food commodities procured for UK consumption through a diversity of supply chains. Final Report for Defra Project FO0103, Defra, UK.

Wood, R., Lenzen, M., Dey, C. & Lundie, S. 2006. A comparative study of some environmental impacts of conventional and organic farming in Australia. *Agricultural Systems*, 89: 324–348.

Press, Cambridge, United Kingdom and New York, NY, USA (available at http://www.ipcc.ch/publications_ and_data/ar4/wg3/en/contents.html).

Thomassen, M.A., van Calker, K.J., Smits, M.C.J., Iepema, G.L. & de Boer, I.J.M. 2008. Life cycle assessment of conventional and organic milk production in the Netherlands. *Agricultural Systems*, 96: 95–107.

van der Werf, H.M.G., Tzilivakis, J., Lewis, K. & Basset-Mens C. 2007. Environmental impacts of farm scenarios according to five assessment methods. *Agriculture Ecosystems and Environment*, 118: 327–338.

van Zeijts, H., Leneman, H. & Sleeswijk A.W. 1999. Fitting fertilisation in LCA: Allocation to crops in a cropping plan. *Journal of Cleaner Production*, 7:69-74.

Voluntary Carbon Standard (VCS) 2010. Quantifying N_2O emissions reductions in US agricultural crops through N-Fertilizer rate reduction. Voluntary Carbon Standard proposed methodology (available athttp://www.v-c-s.org/methodologies/quantifyingn2o-emissions-reductions-us-agricultural-crops-through-n-fertilizer-rate-0).

Wenzel, H., Hauschild, M. & Alting, L. 1997. Environmental assessments of products – Vol. 1: Methodology, tools and case studies in product development. Chapman and hall, London, UK.

Williams, A.G., Audsley, E. & Sandars, D.L. 2006. Determining the environmental burdens and resource use in the production of agricultural and horticultural commodities. In: **Main Report, Defra Research Project IS0205**. Cranfield University and Defra, UK (available at http://www.defra.go.uk).

Williams, A.G., Pell, E., Webb, J., Tribe, E., Evans, D., Moorhouse, E. & Watkiss P. 2008. Comparative life cycle assessment of food commodities procured for UK consumption through a diversity of supply chains. Final Report for Defra Project FO0103, Defra, UK.

Wood, R., Lenzen, M., Dey, C. & Lundie, S. 2006. A comparative study of some environmental impacts of conventional and organic farming in Australia. *Agricultural Systems*, 89: 324–348.

